The modes of origin
of lowest organisms

HENRY CHARLTON BASTIAN

1871

TABLE OF CONTENTS

PREFACE

Having been compelled by the results of my investigations on the question of the Origin of Life to arrive at conclusions adverse to generally received opinions, I found that several persons having high authority in matters of science, were little disposed to assent to these views. To a great extent this seemed due to the fact that a distinguished chemist had previously gone over some of the same ground, and had arrived at precisely opposite conclusions. M. Pasteur has been long known as an able and brilliant experimenter, and some of his admirers seem to regard him as an almost equally faultless reasoner.

Renewed and prolonged experimentation having tended to demonstrate the truth of my original conclusions, and to convince me of the utter untenability of M. Pasteur's views, it seemed that the best course to pursue would be, at first, to endeavour to show into what errors of reasoning M. Pasteur had fallen, and also how his conclusions were capable of being reversed by the employment of different experimental materials, and different experimental methods. Then, having presented, in a connected form, evidence which might suffice to shake the faith of all who preserved a right of independent judgment, one might hope to have paved the way for the reception of new views—even though they were adverse to those of M. Pasteur. The present volume contains, indeed, only a fragment of the evidence which will be embodied in a much larger work—now almost completed—relating to the nature and origin of living matter, and in favour of what is termed the physical doctrine of Life.

The question of the mode of origin of Living Matter, is inextricably mixed up with another problem as to the cause of fermentation and putrefaction.

M. Pasteur's labours were, at first, undertaken in order to solve the latter difficulty—to decide, in fact, between two rival hypotheses. It was held, on the one hand, that many ferments were mere dead nitrogenous substances, and that fermentation was a purely chemical process, for the initiation of which the action of living organisms was not necessary; whilst, on the other hand, it was also maintained that no fermentation could be initiated without the agency of living things—in fact, that all ferments were living organisms. The former may be called the physical theory of fermentation, of which Baron Liebig is the most prominent modern exponent; whilst the latter may be termed the vital theory of fermentation, and this is the doctrine of M. Pasteur. All the facts which I have to adduce, so far as the subject of fermentation is concerned, are wholly in favour of the views of Baron Liebig.

And, the conclusions arrived at in this work are confirmed by the results of several unpublished experiments, in which living organisms have been taken from flasks that had, a few weeks before, been hermetically sealed and heated for a variable time to temperatures ranging from 260° F. to 302° F.

With the view of aiding some of my readers in their interpretation of the results of some of the experiments contained in this volume, I would call their attention to the following considerations. If fluids in vacuo (in hermetically-sealed flasks), which were clear at first, have gradually become turbid; and if on microscopical examination this turbidity is found to be almost wholly due to the presence of Bacteria or other organisms, then it would be sheer trifling gravely to discuss whether the organisms were living or dead, on the strength of the mere activity or languor of the movements which they may be seen to display. Can dead organisms multiply in a closed flask to such an extent as to make an originally clear fluid become quite turbid in the course of two or three days?

And if any one wishes to convince himself as to whether such turbidity can occur in a flask which is still hermetically sealed, let him take one that has been prepared in the manner I have elsewhere described, carefully heat the neck of it in a spirit-lamp flame, and see how the rapid in-bending of the red-hot glass testifies to the preservation of a partial vacuum within. The vacuum in such cases is only partially preserved, because of the emission of a certain amount of gases within the flask—such as invariably occurs during the progress of fermentation or putrefaction.

In these experiments with heated fluids in closed flasks, nothing is easier than to obtain negative results. The same kinds of infusions which—if care has been taken to obtain them strong enough—will in a few days teem with living organisms, often show no trace of living things after much longer periods, when the solutions are weak. Again, in those cases where only a few organisms exist in a solution which has been made the subject of experimentation, nothing is easier than by a perfunctory examination of the fluid to fail in finding any of these sparsely-distributed living organisms. Experiments, the results of which are positive, may, therefore, in the absence of sufficient care, be cited as negative; and experiments which would otherwise have been crowned with unmistakeably positive results, may be rendered wholly barren by the employment of infusions which have been carelessly made.

A word of explanation seems necessary with regard to the introduction of the new term Archebiosis. I had originally, in unpublished writings, adopted the word Biogenesis to express the same meaning—viz., life-origination or commencement. But in the mean time the word Biogenesis has been made use of, quite independently, by a distinguished biologist, who wished to make it bear a totally different meaning. He also introduced the word Abiogenesis. I have been informed, however, on the best authority, that neither of these words can—with any regard to the language from which they are derived—be supposed to bear the meanings which have of late been publicly assigned to them. Wishing to avoid all needless confusion, I therefore renounced the use of the word Biogenesis, and being, for the reason just given, unable to adopt the other term, I was compelled to introduce a new word, in order to designate the process by which living matter is supposed to come into being, independently of pre-existing living matter.

H. Charlton Bastian.
Queen Anne Street, W.,
May 8, 1871.

THE MODES OF ORIGIN OF LOWEST ORGANISMS

The mode of origin of Bacteria, and, to a less extent, of Torulæ, has been much discussed of late, and many different views have been advocated on this subject by successive writers.

It is of much importance to bear in mind when such views are under consideration, that a short time since nothing was positively known concerning the life-history of these organisms. However strongly, therefore, certain persons are inclined to rely upon the analogy which is supposed to obtain between these doubtful cases, and the multitudes of known cases— in which it can be shown that organisms are the offspring of pre-existing organisms—it must always be borne in mind that in many of the doubtful cases, where the simplest organisms are concerned, there is also an analogical argument of almost equal weight adducible in favour of their de novo origination—after a fashion, and under the influence of laws similar to those by which crystals arise. To rely too exclusively upon an argument from analogy is always perilous: it is more than usually so, however, in a case like this, where what is practically an opposing analogy may be deemed to speak just as authoritatively in an opposite direction.

There is one consideration, moreover, which deserves to be pointed out here, and which does not seem to have occurred to most of those who so firmly pin their faith to the truth of the motto "omne vivum ex vivo." The every-day experience of mankind, supplemented by the ordinary observations of skilled naturalists, does pretty fairly entitle us to arrive at a wide generalization, to the effect that some representatives of every kind of organism are capable of reproducing similar organisms. But, whilst this is all that the actual every-day experience of mankind warrants being said, and whilst there is in reality the widest possible gulf between such a generalization and that which is expressed by the motto "omne vivum ex vivo," the latter formula has of late been spoken of as though it were the one which was in accordance with the daily experience of mankind, rather than the other, which gives expression to a generalization of a much narrower description. This experience, in reality, affords no evidence which could entitle us to place implicit belief in the formula "omne vivum ex vivo."

Whilst we do know something about the ability which most organisms possess of reproducing similar organisms, we cannot possibly say, from direct observation, that every organism which exists has had a similar mode of origin, because the cases in which organisms may have originated de novo are the very cases in which their mode of origin must elude our actual observation. Such a statement, too, would be all the more dangerous, in the face of the other analogy, when it can actually be shown that some organisms do make their appearance in fluids after precisely the same fashion as crystals.

Although, therefore, there is a contradiction between the unwarrantable and ill-begotten formula, "omne vivum ex vivo," and the doctrines of what has been called "Spontaneous Generation"; there is no contradiction whatever between such doctrines and the only generalization which we are really warranted in arriving at, to the effect that some representatives of every kind of organism are capable of reproducing similar organisms.

Bacteria, Torulæ, or other living things which may have been evolved de novo, when so evolved, multiply and reproduce just as freely as organisms that have been derived from parents.

The views as to the origin of Bacteria and Torulæ which are most worthy of attention, may be thus enumerated:—

a. That they are independent organisms derived by fission or gemmation from pre-existing Bacteria and Torulæ.

b. That they represent subordinate stages in the life-history of other organisms (fungi), from some portion of which they have derived their origin, and into which they again tend to develop.

c. That they may have a heterogeneous mode of origin, owing to the more complete individualization of minute particles of living matter entering into the composition of higher organisms, both animal and vegetal.

d. That they may arise de novo in certain fluids containing organic matter, independently of pre-existing living things (Archebiosis).

I shall make some remarks concerning each of these views, though the evidence I have to adduce mainly concerns the possibility of the origin of Bacteria and Torulæ in the way last alluded to, viz., by Archebiosis.

The third mode of origin is what is called Heterogenesis; whilst the first and second modes are the representatives of more familiar processes, included under the head of Homogenesis. Thus, in accordance with the first view,

Bacteria may be regarded as low organisms having a distinct individuality of their own and multiplying by a process of fission—thus affording instances of what I propose to term direct Homogenesis. Whilst, in accordance with the second view, Bacteria are supposed to represent merely one stage in the life-history of higher organisms, which are therefore reproduced by an indirect or cyclical process of Homogenesis.

The possible modes of origin of Bacteria and Torulæ may, therefore, be tabulated as follows:—

Modes of origin
of Bacteria and
Torulæ. . Homogenesis. a. Direct.
b. Indirect.
. Heterogenesis.
. Archebiosis.

I. Homogenetic Mode of Origin of Bacteria and Torulæ.

Bacteria and Torulæ being already in existence, they may, undoubtedly, reproduce organisms similar to themselves by processes of fission and gemmation—in the same way that other low protistic organisms propagate their kind. Although so many reasons rendered this view probable, it was some time before I was able actually to confirm it by personal observations in the case of Bacteria. In the ordinary microscopical examination of portions of an infusion containing these organisms, an observer may watch for hours and never see a single instance of such fission occurring. His attention is apt to be distracted by the number of organisms which are constantly flitting before his view, and he is, moreover, perhaps apt to pay particular attention to those which seem by their movements to be most obviously alive.

I have observed the process most plainly when a few Bacteria have been enclosed in a single drop of fluid, pressed into a very thin stratum, in a "live-box" kept at a temperature of about ° Fahr. by resting on one of Stricker's warm-water chambers placed on the stage of the microscope. Under these conditions, I have seen a Bacterium of moderate size divide into two, and each of these into two others somewhat smaller, in the course of fifteen minutes.

It is still more worthy of remark, that in all cases (so far as I have been able to observe), this, the most certain sign of vitality which such organisms are capable of manifesting, is shown by those which, from their stillness, might be considered dead. The Bacteria which are about to divide are generally either motionless, or merely present slight oscillating movements. The separation is quickly brought about at the joint, so that the original organism divides into two equal portions; and these, lying close together, soon develop a new construction as they grow, through which a further division may occur.

That the Bacteria which reproduce should be in a comparatively quiescent condition, seems not difficult to understand. Such rudimentary organisms do not appear to possess cilia or other locomotory appendages: their movements are, therefore, in all probability dependent upon the mere molecular changes which are taking place within them, and upon which their life and nutrition depend. The process of fission must, however, be considered as the result of a new effort at equilibrium, which has, perhaps, been necessitated by molecular changes that have occurred during a preceding period of growth. The living matter which is no longer able to exist round a single centre, re-arranges itself around two centres,—as a result of which, fission occurs. It seems only natural, therefore, that whilst this active work of molecular re-arrangement is going on, those other molecular movements which occasion the actual locomotion of the organism from place to place, should be more or less interfered with.

This is the one and only mode of multiplication of Bacteria and of Torulæ which is actually known to occur; and such a limitation is in accordance with the more general fact, that processes of fission or gemmation are the only means of reproduction that are known to occur in the lower kinds of organisms, belonging to the PROTISTIC kingdom.

However well this process of fission may have been established, as a frequent mode of reproduction of Bacteria, such a fact does not lend any support to the notion that these are necessarily distinct and independent organisms. Torulæ (of which beer-yeast is the most familiar example) may similarly undergo this process of mere vegetative repetition to an indefinite extent, whilst only some of the products develop into fungi. The gonidia of lichens may also reproduce indefinitely in this fashion, and only some of the products of multiplication may go on to the production of lichens similar to that from which the gonidia had been derived.

It is a fact, however, admitted by many, and which any patient microscopist is capable of verifying for himself, that some Bacteria do develop into Leptothrix filaments, and that these are capable of passing into a dissepimented mycelial structure of larger size and undoubtedly fungus nature—from which fructification of various kinds may be produced. Some Bacteria may therefore develop into some fungi, just as certainly as some Torulæ may develop into other fungi, or, just as surely as some multiplying gonidia may develop into lichens.

In order to prove, however, that the Bacteria which happen to go through this development into Leptothrix and thence into fungi, are strictly to be considered as necessary links in the life-history of fungi, it would be essential for the person holding such views, to show that Bacteria could not arise independently—or at least that no independently evolved Bacteria could develop through Leptothrix-forms into a fungus. And, similarly, for the other kinds of organisms: in order to establish that the Torula cell is a necessary link in the life-history of certain fungi, or the gonidial cell a necessary link in the life-history of lichens, it would be necessary to show that Torulæ or gonidial cells could not originate de novo—that no independently evolved Torula or gonidial cell could develop into a fungus or a lichen.

An easier position to establish would be, that the Bacterium or the Torula were occasionally links in the life-history of fungi, or that the gonidial cell was an occasional link in the life-history of a lichen. This doctrine would leave the other more difficult problems,—as to the possible existence of supplementary modes of origin for such organisms by Heterogenesis or by Archebiosis—perfectly open questions.

To establish the position that Bacteria are occasional links in the life-history of fungi, it would be only necessary to show that some of the Bacteria which develop into fungi through Leptothrix have derived their origin from pre-existing fungi. This is the view which Hallier has endeavoured to establish; it is also the doctrine of M. Polotebnow, and one, moreover, to which Professor Huxley inclines. Even this mode of origin for Bacteria, however, has not been so decisively established as might be desired. With regard to Torulæ, we do possess sufficient evidence tending to show that some of them may arise from pre-existing fungi, and we are equally certain that some gonidial cells are thrown off from lichens. The analogical evidence is, therefore, in favour of the view that minute particles which are budded off from the mycelium of certain fungi, may subsequently lead an independent existence, and multiply in the form of Bacteria—although many of the cases in which such buds seem to be given off, may be merely cases in which co-existing Bacteria have become adherent to fungus filaments or to Torulæ.

But, with reference to these supposed cases of budding, and also to those others in which the contents of a spore or sporangium break up into what Professor Hallier calls "micrococci" (which are generally incipient Bacteria), it would be difficult for us to decide whether such processes are normal or abnormal. When we have to do with such organisms, in fact, there may be the nicest transitions between what is called Homogenesis, and what, when occurring in other organisms, we term Heterogenesis. It may be that the production of such "micrococci" from the spore or sporangium of the fungus is not an invariable incident in the life-history of the species, but rather an occasional result of the influence of unusual conditions, or of failing vigour on the part of the organism. In this latter case we should have to do with a process of Heterogenesis; although, as I have just stated, in respect to such low and changeable organisms, scarcely any distinct line of demarcation can be drawn between Homogenesis and Heterogenesis.

The evidence seems, therefore, against the notion that Bacteria or Torulæ are ordinary, independent living things, which merely reproduce their like.

That some Bacteria are produced from pre-existing Bacteria, just as some Torulæ are derived from pre-existing Torulæ, may, it is true, be considered as settled. But, so far as we have yet considered the subject, there may be just as good evidence to show that Bacteria and Torulæ are capable of arising de novo, as there is that some of them are capable of developing into fungi.

If this were the case, such types could only be regarded as the most common forms assumed by new-born specks of living matter; and, by reason of their origin—which would entail an absence of all hereditary predisposition—they might be supposed to be capable of assuming higher developmental forms.

Now, as a matter of fact, worthy of arresting our attention, we do find that some Bacteria are capable of growing into Leptothrix, whilst this is able to develop continuously into a fungus; just as we also know that some Torulæ are capable of growing into other fungi.

Should it be established, therefore, that Bacteria and Torulæ are capable of arising de novo, the facts concerning their mutability are harmonious enough with theoretical indications.

But, as I have before indicated, although it is quite true that some Bacteria develop into fungi, such forms may constitute no necessary links in the life-history of other fungi. I have suggested that in those (occasional) cases in which they do occur as links in the life-history of fungi, there is room for doubt whether these Bacteria are to be considered as normal products, or as abnormal results (heterogeneous offcasts), brought about by some unusual conditions acting upon the parent fungus. That is to say, we may be doubtful whether in such a case their origin ought to be considered Homogenetic or Heterogenetic. It may be that many of the lower fungi are such changeable organisms, and so prone to respond to the various "conditions" acting upon them (which would be almost certainly the case if they had been developed from a Bacterium in two or three days—the Bacterium itself having been evolved de novo) that no very valid distinction can here be drawn between Homogenesis and Heterogenesis. Our whole point of view, in fact, concerning such fungi as are seen to develop through Leptothrix forms from Bacteria must be entirely altered, if it is once conceded that Bacteria may arise de novo. Such simple Mucedineæ would then have to be regarded as mere upstart organisms only a few removes from dead matter, and—in view of the greater molecular mobility of living matter—capable of being modified in shape and form even more than the most changeable crystals under the influence of altering "conditions." We should have no longer to do with the members of a stable species, which had been reproducing its like through countless geologic ages anterior to the advent of man upon the earth. Indeed, in order to reconcile such a possibility with the seemingly contradictory fact of the known extreme changeability of these lower forms of life, we hear only vague hints thrown out about our imperfect knowledge of the "limits within which species may vary." As if, in the face of what we do know concerning hereditary transmission, this changeability did not make it almost impossible to conceive that there should have been an unbroken series of such organisms since that remote epoch of the earth's history, when the first organisms of the kind made their appearance. It does not seem to me that the presumed permanence of a very changeable organism is consistent with, or rendered more explicable by, the supposition that some representatives of the species have constantly been undergoing progressive modifications which have been successively perpetuated by inheritance, in the shape of distinct specific forms. Why should some be presumed to have undergone so much change, whilst others (presenting an equal and an extreme degree of modifiability, even to the present day) are supposed to have preserved the same specific form through a countless series of changing influences?

. Heterogenetic Mode of Origin of Bacteria and of Torulæ.

It has been long known that Bacteria and Torulæ are frequently to be found within vegetable cells, taken even from the central parts of plants, whenever these are in a sickly condition or are actually dying. They are apt to exist also within epithelial cells taken from the inside of the mouth; and the frequency and abundance with which such organisms are met with in these cells, is almost in direct proportion to the malnutrition and lack of vital power in the individual who is the subject of observation. Then, again, in persons who have died of adynamic diseases, in the course of twenty-four or thirty-six hours (during warm weather) Bacteria may be found in abundance within the blood-vessels of the brain and of other parts, although no such Bacteria were recognizable in the blood of the individual during life.

In these cases we must, in order to account for the presence of the Bacteria and Torulæ, either suppose that such organisms, in an embryonic state, are almost universally disseminated throughout the various textures of higher organisms, both animal and vegetal (though they are only able to develop and manifest themselves when the higher organisms, or the parts of them in which the Bacteria or Torulæ are met with, are on the eve of death), or else we must imagine that when the vital activity of any organism, whether simple or complex, is on the wane, its constituent particles (being still portions of living matter) are capable of individualizing themselves, and of growing into the low organisms in question. Just as the life of one of the cells of a higher organism may continue for some time after the death of the organism itself, so, in accordance with this latter view, may one of the particles of such a cell be supposed to continue to live after even cell-life is impossible.

Now, to many persons, the latter seems to be a much simpler hypothesis than the former, and one, moreover, which is more in accordance with known facts. People's views, however, on this subject are likely to be much influenced by their notions as to the possibility of Bacteria arising by a process of Archebiosis. Although some may be inclined to accept the doctrine of Heterogenesis, the same persons, being "vitalists," may not readily believe in the doctrine of Archebiosis, because this implies the vivification of dead matter—the conversion of not-living elements into a living combination. Those, however, who do believe in Archebiosis will—if the necessary evidence be forthcoming—all the more readily yield their assent to the doctrine of Heterogenesis, because it is a much less novel thing to have to believe in the mere transformation of living matter, than in the possibility of its origin de novo.

Evidence of a tolerably satisfactory nature, however, is forthcoming, which may speak independently in favour of the doctrine of Heterogenesis.

It has been affirmed by Crivelli and Maggi that they have actually seen the particles within granular epithelial cells (taken from the back of the tongue of a patient suffering from diabetes) grow and elongate, so as to give rise to Bacteria, or fuse in longitudinal series, so as to form a Vibrio. And, moreover, as I have myself ascertained, if one takes healthy-looking epithelial scales scraped from the inside of the mouth, which appear to contain nothing but the finest granules, and places them with a little saliva in a "live-box" (and this within a damp chamber kept at a temperature of about ° Fahr.), in the course of from to hours, the cells may be found to be studded throughout with motionless Bacteria. Of course it may be said that the granules originally seen in the cells were offcasts from pre-existing Bacteria which had gained access to the cell. And although, to many, this may seem an extremely improbable supposition, it is, nevertheless, one which it would be very difficult to disprove. The improbability of the notion is increased, moreover, when we find that Bacteria, and even Torulæ, will develop just as freely within closed cells taken from the very centre of a vegetable tuber, as they will in the midst of the more solid epithelial cell from the inside of the mouth. If it be urged that in this latter situation, there is the greatest chance of the cells being brought into contact with Bacteria, and that it must be considered possible for imaginary minute offcasts from these Bacteria to make their own way into the substance of the epithelial cell, I am quite willing to grant the desirability of taking such possibilities into consideration. But, at the same time, it seems all the less likely that the actual occurrence of the Bacteria is explicable on these grounds, because we find them developing just as freely within the cells freshly cut from the centre of a tuberous root, or we may find them already developed within these cells, if the root has begun to decay. To suppose that actual germs of Bacteria and of Torulæ are uniformly distributed throughout the tissues of higher organisms, is to harbour a hypothesis which would appear to many to be devoid of all probability—more especially when the heterogenetic mode of origin of larger and higher organisms is a matter of absolute certainty.

. Origin of Bacteria and of Torulæ by Archebiosis.

The evidence on this part of the subject is, I think, sharply defined and conclusive. Simple experiments can be had recourse to, which are not admissible in the discussion of the question as to the origin of Bacteria and Torulæ by Heterogenesis. There, we wish to establish the fact that living matter is capable of undergoing a certain metamorphosis, and consequently, we must deal with living matter. Here, however, with the view of establishing the fact that living matter can arise de novo, if we are able, shortly after beginning our experiment, to arrive at a reasonable and well-based assurance that no living thing exists in the hermetically sealed experimental vessel—if the measures that we have adopted fully entitle us to believe that all living things which may have pre-existed therein have been killed—we may feel pretty sure that any living organisms which are subsequently found, when the vessel is broken, must have originated from some re-arrangements which had taken place amongst the not-living constituents of the experimental solutions, whereby life-initiating combinations had been formed.

The possibility of the de novo origination of Bacteria, Torulæ, and other such organisms, is one which is intimately associated with the doctrine as to the cause of fermentation and putrefaction. With regard to the almost invariable association of such organisms with some of these processes, almost all are agreed. There is, moreover, a very frequent association of particular kinds of organisms with particular kinds of fermentation. Hence the assumption is an easy and a natural one to many persons, that the organisms which are invariably met with in some cases are the causes of these fermentations, although it is quite obvious that the facts on which this view is based, are equally explicable on the supposition that the organisms are concomitant results or products (due to new chemical combinations) of the fermentative changes. In the one case the fermentative changes are believed to be initiated by the influence of living organisms; and those who regard living things as the only true ferments, for the most part also believe that living things are incapable of arising de novo. They think that those organisms which serve to initiate the changes in question, have been derived from a multitudinous army of omnipresent atmospheric germs, which are always ready, in number and kind suitable for every emergency. This is the doctrine of M. Pasteur and others. On the other hand, fermentations and putrefactions may be regarded as sets of chemical changes, which are apt to occur in organic and other complex substances—these changes being due either to the intrinsic instability of the body which manifests them, or to molecular movements communicated to it by a still more unstable body. Baron Liebig says:—"Many organic compounds are known, which undergo, in presence of water, alteration and metamorphosis, having a certain duration, and ultimately terminating in putrefaction; while other organic substances that are not liable to such alteration by themselves, nevertheless, suffer a similar displacement or separation of their molecules, when brought into contact with the ferments."

Each substance belonging to the first class, would be at the same time, therefore, both ferment and fermentable substance; whilst a small portion of such substance, when brought into contact with a less unstable substance, might induce such molecular movements as to make it undergo a process of fermentation. With regard to the cause of such induced fermentative changes, Gerhardt says, in explaining Liebig's views:—"Every substance which decomposes or enters into combination is in a state of movement, its molecules being agitated; but since friction, shock, mechanical agitation, suffice to provoke the decomposition of many substances (chlorous acid, chloride of nitrogen, fulminating silver), there is all the more reason why a chemical decomposition in which the molecular agitation is more complete, should produce similar effects upon certain substances. In addition, bodies are known which when alone are not decomposed by certain agents, but which are attacked, when they exist in contact with other bodies incapable of resisting the influence of these agents. Thus platinum alone does not dissolve in nitric acid, but when allied with silver, it is easily dissolved; pure copper is not dissolved by sulphuric acid, but it does dissolve in this when it is allied with zinc, andc. According to M. Liebig it is the same with ferments and fermentable substances; sugar, which does not change when it is quite alone, changes—that is to say ferments—when it is in contact with a nitrogenous substance undergoing change, that is, with a ferment."

Thus, in accordance with this latter view, living ferments are not needed— mere dead, organic or nitrogenous matter suffices to initiate the processes in question. Those who hold this opinion may or may not believe that organisms are capable of arising de novo; though there can be little doubt that a belief in the truth of such a doctrine does, almost inevitably, entail a belief in the de novo origination of living things. No one who has looked into the evidence, doubts the fact of the association between some of these processes and the presence of organisms; the only question is, as to the relation in which they stand to one another. If organisms are not the causes of those fermentative changes with which they are invariably associated, then they are, in all probability, the results of such changes; and they must certainly have been produced de novo if it can be shown that fermentation or putrefaction may take place under the influence of conditions which make it certain that pre-existing living organisms could have had nothing to do with the process.

Now, in order to lend some air of probability to the former hypothesis, concerning the necessity for the existence of living ferments, it was incumbent upon its supporters to endeavour to show that the air did contain such a multitude of "germs," or living things, as were demanded by the requirements of their theory. Spallanzani and Bonnet had, as far as the imagination was concerned, done all that was necessary. They had proclaimed the universal diffusion of "germs" of all kinds of organisms throughout the atmosphere—which were ready to develop, whenever suitable conditions presented themselves. So far, however, this was but another hypothesis. To establish the doctrine that fermentation cannot take place without the agency of living ferments, we cannot receive hypotheses in evidence: facts are needed. These, no one attempted to supply in an adequate manner anterior to the investigations of M. Pasteur. Speaking of his researches, even M. Milne-Edwards says, "Previous to this time, the existence of reproductive particles, or infusorial germs in the atmosphere was nothing more than a plausible hypothesis, put forward in order to explain the origin of such creatures in a manner conformable with the general laws of reproduction; but it was only a mere supposition, and no one had been able actually to see or to handle these reproductive corpuscles."

We have to look, therefore, to M. Pasteur's investigations, and to others which may have been since conducted, for all the scientific evidence in support of what has been called the "Panspermic hypothesis."

By an ingenious method of filtration, which is fully described in his memoir, M. Pasteur separated from the air that passed through his apparatus the solid particles which it contained. This search convinced him that there were, as he says, "constantly in ordinary air a variable number of corpuscles whose form and structure declare them to be organized." Some of these, he thinks, resemble the spores of fungi, and others the ova of ciliated infusoria, though he adds:—"But as to affirming that this is a spore, much less the spore of any definite species, and that one is an egg, and belonging to such an infusorium, I believe that this is not possible." He limits himself, in fact, to the statements, that the corpuscles which he found, were (in his opinion) evidently organized; that they resembled in form and appearance the germs of the lower kinds of organisms; and that, from their variety in size, they probably belonged to many different sorts of living things. Even here, therefore, we have to do with the impressions of M. Pasteur, rather than with verified statements. All that has been established by his direct investigation as to the nature of the solid bodies contained in the atmosphere is this: that the air contains a number of round or ovoidal corpuscles, often quite structureless, which he could not distinguish from the spores of fungi—some of which, being about the right size, were round or ovoidal, and structureless. In addition, however, it has been shown that the air contains other rounded corpuscles which are similarly structureless, though composed of silica or starch. It may therefore be asked, in the first place, whether the conclusion is a sufficiently safe one that many of the corpuscles found by M. Pasteur were spores of fungi; and in the next place, supposing this to have been established, whether such spores were living or dead. These questions would have been answered satisfactorily if M. Pasteur could state that he had actually watched the development of such corpuscles, in some suitable apparatus, into distinct organisms. But any such development, he distinctly states, he never witnessed. He says:— "What would have been the better and more direct course would have been to follow the development of these germs with the microscope. Such was my intention; but the apparatus which I had devised for this purpose not having been delivered to me at a convenient time, I was diverted from this investigation by other work." The evidence which he does adduce, in subsequent portions of his memoir, in order to prove that some of these corpuscles were really "fertile germs," is almost valueless, because all the facts are open to another interpretation, which is just as much, nay, even more, in accordance with Baron Liebig's than with his own doctrine of fermentation.

But another most important consideration presents itself. M. Pasteur's researches as to the nature of the dust contained in the atmosphere enable him to say nothing concerning the presence of Bacteria, although he himself admits that these are generally the first organisms which display themselves in fermentations or putrefactions, and that in a very large majority of the cases in which fermentation occurs in closed vessels they are the only organisms which make their appearance. And yet, notwithstanding these facts, M. Pasteur says, in reference to the common form of Bacterium:—"This infusorial animal is so small that one cannot distinguish its germ, and still less fix upon the presence of this germ, if it were known, amongst the organized corpuscles of the dust which is suspended in the air."

Here, then, we have a confession from M. Pasteur himself, that all evidence fails, where it is most wanted, in support of his hypothesis.

If a large number of fermentations begin with the presence of Bacteria as the only living things, and if in a number of cases no other organisms ever occur, it is useless to adduce as evidence, in proof of the view that fermentations are always initiated by air-derived organisms, the fact that certain corpuscles (supposed to be spores of fungi) are recognizable in the atmosphere—capped by the distinct statement that Bacteria or their germs are not recognizable. If Bacteria are not recognizable in the atmosphere, what scientific evidence is there that the fermentations in which these alone occur are initiated by Bacteria derived from the atmosphere, or from certain imaginary Bacteria germs, which we are supposed to be unable to distinguish? M. Pasteur may, moreover, be reminded that when he resorts to the supposition of Bacteria possessing "germs" which are indistinguishable, he is again resorting to hypothesis rather than to fact, in order to prove the truth of the particular doctrine of fermentation which he advocates. Bacteria are known to reproduce and multiply only by a process of fission; each of the parts into which they divide being nothing more than a part of the original Bacterium, and therefore endowed with similar properties of resisting heat, desiccation, and other agencies. Any resort to invisible germs to account for the multiplication of Bacteria, which are known to reproduce freely in other ways, is obviously not permissible, unless such postulation be more or less necessitated by the occurrence of facts otherwise inexplicable.

Although, therefore, no direct evidence has been adduced tending to show that Bacteria are present in the atmosphere, even if this evidence had been forthcoming, it would have been necessary, in reference to M. Pasteur's hypothesis, for it to be supplemented by further evidence to the effect that Bacteria were well capable of resisting such an amount of desiccation as must have been involved by their presence for an indefinite time in the atmosphere even of the hottest and driest regions of the earth. For, organic substances in solution do not only putrefy in moist weather or moist climates; they putrefy most rapidly and surely when the temperature is high, and quite irrespectively of the amount of moisture contained in the atmosphere. A property of resisting the effects of desiccation—the possession of which, by Bacteria, is so necessary for the truth of M. Pasteur's argument—ought to have been shown by scientific evidence to be a real attribute of such organisms; though it seems, on the contrary, to have been assumed to exist, with almost equal readiness by both parties, in the controversies concerning the possibility of "spontaneous generation." This error may be ascribed to the misguiding influence of a treacherous analogy. Whilst it may be true that certain seeds and spores, and also that Rotifers, "Sloths," and some Nematoids are capable of resisting the influence of a prolonged exposure to desiccating influences, it may well be asked, whether the same fact necessarily holds good for organisms such as Bacteria, which have no chitinous or other envelopes to protect them, and which are merely minute fragments of naked protoplasm. Having elsewhere shown how far presumptions had stolen a march upon established facts, in reference to the supposed possession of a similar property by the Free Nematoids, my eyes were opened to the reality of this uncertainty with regard to Bacteria. It is, however, no easy matter definitely to prove or to disprove the possession of this property by organisms so minute as Bacteria, and therefore so difficult to identify. If dried Bacteria are added to a drop of a suitable solution—similar to that in which they had been bred—it soon becomes quite impossible to distinguish those which have been added from those which arise in the fluid. Taking into consideration the fate of other simple organisms, however, it is by no means improbable that they should be killed even by a short desiccation. I have found, for instance, that desiccation for half-an-hour in a room at a temperature of ° F. suffices to kill all the larger, naked, lower organisms with which I have ex-peri-men-ted—in-clud-ing long Vibrios, Amœbæ, Monads, Chlamydomonads, Euglenæ, Desmids, Vorticellæ and all other Ciliated Infusoria.

But, certain indirect evidence seems to speak most authoritatively against the supposition that the air contains any notable quantity of living Bacteria, or Bacteria germs, whether visible or invisible. I have always found that a simple solution of ammonic tartrate, which has been placed—without previous boiling—in a corked bottle of greater capacity, will become turbid in two or three days, owing to the presence of myriads of Bacteria; whilst a similar solution, previously boiled, may remain for ten days, three weeks, or more, without showing the least trace of turbidity, although the open neck of the bottle or flask in which it is contained, may be covered only by a loose cap of paper. And yet, at any time, in order to make this fluid become turbid in from to hours, all that one has to do is to bring it into contact with a small glass rod which has just been dipped into a solution containing living Bacteria.

If we find that an eminently inoculable fluid will remain for two or three weeks, or perhaps more, in contact with the air without becoming turbid, though it will always become turbid in two or three days if brought into contact with living Bacteria, what can we conclude, but that living Bacteria are not very common in the atmosphere? These most striking facts can be easily verified by other observers.

Thus we find ourselves, at present, in this position. After all that has been said and done to prove the wonderful prevalence of "germs" in the atmosphere, we are really still in the region of hypothesis—no further advanced than we were in the time of Bonnet and of Spallanzani, so far as it concerns the organisms which are all important—Bacteria. Neither these nor their germs have been shown to exist in any recognizable abundance in the atmosphere, and yet in most fermentations they are the first organisms which display themselves; whilst in many such fermentations Bacteria alone occur. Nay more, even were they present in any great abundance, there is some reason to believe that the majority of them would exist as mere dead, organic particles—because Bacteria are more than likely to be unable to resist anything like an extreme or prolonged exposure to desiccating influences.

The first and essential data in support of M. Pasteur's hypothesis must, therefore, be regarded as entirely unproved in respect to Bacteria—which are the most important of all organisms, in relation to the cause of fermentation and putrefaction.

Without the aid of elaborate experiments, however, the evidence which the microscope can supply is capable of leading us to the conclusion that such search for atmospheric Bacteria germs, was comparatively useless. If it can be shown that Bacteria can arise in a fluid independently of visible germs, then, obviously, any inquiries as to the nature of the visible contents of the atmosphere, can have only a very indirect bearing upon the question as to the mode of origin of these organisms. And yet by the aid of the microscope, as I have elsewhere stated, one can watch the appearance of almost motionless specks, more or less uniformly diffused through a motionless film of fluid, and can see them gradually develop into moving Bacteria or into Torulæ. So that, where no visible germs previously existed, visible particles of living matter develop, and more or less rapidly grow into distinct Bacteria. This may be best seen in a drop of a fresh and very strong turnip infusion, which has been filtered several times through the finest paper. The drop, placed in a live-box, should be flattened into a thin film by the application of the cover.

Thus protected, evaporation takes place very slowly, and with the live-box resting on one of Stricker's hot-water plates, at a temperature of ° to ° F., and the latter upon the stage of the microscope, one can easily select a portion of the field in which either no particles or only a countable number exist. If, therefore, around and between any mere granules which may pre-exist, or in a clear space, one gradually sees in the course of two or perhaps three hours, a multitude of almost motionless specks (at first about " in diameter) in positions where no such specks previously existed; and if these specks may be seen gradually to increase in size and develop into Bacteria and Torulæ, then, at all events, we are able to say that these organisms can be developed without pre-existing visible germs, and we have just the same amount of actual evidence for believing that they have been formed de novo, as we should have for believing that crystals had been formed de novo, if we had seen them appearing under our eyes in the same manner. Whether they really arise after the fashion of crystals, without the aid of pre-existing though invisible germs, is a matter which can only be settled inferentially, by a subsequent resort to strict methods of experimentation.

Seeing however, that we are able, with the aid of the microscope alone, to demonstrate that Bacteria and Torulæ can develop in situations where no visible germs had previously existed, it is useless, as I have said before—so far as the question of their mode of origin is concerned—to search the atmosphere to ascertain what visible germs it may contain. If some Bacteria and Torulæ arise from germs at all, it must be from germs which are invisible to us. The finding of visible germs in the atmosphere can, therefore, only have an indirect bearing upon the solution of the problem. Since it can be shown that some visible spores and ova exist in the atmosphere, this affords a certain amount of warrant for the supposition that invisible, living, reproductive particles may also exist—more especially if the existence of an amount of organic matter, which is ordinarily invisible, can be revealed in the air, by the agency of the electric beam, or by any other means.

Nothing can be more illegitimate, however, in the way of inference, than the assumption at once indulged in by Prof. Tyndall and others (who might have been expected, by their previous scientific work, to have learned more caution) that this impalpable organic dust was largely composed of impalpable germs. Yet, without a shadow of proof, without even an attempt to prove it, the air was for a time represented to be a mere stirabout, thick with invisible germs. The briefest reflection, however, upon the probabilities of the case, should have sufficed to suggest a totally different interpretation. The surface of the earth is clothed with living things of all kinds, animal and vegetal, which are not only continually throwing off organic particles and fragments during their life, but are constantly undergoing processes of decay and molecular disintegration after their death. The actual reproductive elements of these living things are extremely small in bulk, when compared with the other parts which are not reproductive, and although Bacteria and Torulæ do exist abundantly, and do materially help to bring about some of the decay in question, yet their bulk, also, is extremely small in comparison with the amount of organic matter itself that is continually undergoing disintegration of a dry kind, in which Bacteria and Torulæ take no part. When, moreover, it is considered that in the neighbourhood of populous cities (the air of which alone exhibits this very large quantity of impalpable, mixed with palpable, organic dust), there is constantly going on a wear and tear of the textile fabrics and of the organic products of various kinds which are daily subservient to the wants of man; and that the chimneys of manufactories and dwelling-houses are also continually emitting clouds of smoke thick with imperfectly consumed organic particles, some idea may be gained of the manifold sources whence the organic particles and fragments found in the atmosphere may emanate, and also as to what proportion of them is likely to be composed of living or dead reproductive elements, or "germs."

Thus, then, so far as the two rival doctrines of fermentation are concerned, the investigation of the nature of the solid particles contained in the atmosphere has revealed facts which are thoroughly in harmony with all the requirements of Liebig's physical theory, though it has almost utterly failed to give anything like a scientific basis to the vital theory of Pasteur. So far from being able to show that living Bacteria (which are the first and oftentimes the only organisms concerned in many processes of fermentation and putrefaction) are universally diffused through the air, Pasteur admits that these cannot be detected, and that their "germs" are not recognizable.

If, therefore, M. Pasteur still maintains the truth of his theory, it should be distinctly understood that it rests originally, not upon established facts, but upon a mere hypothesis—the hypothesis that the air teems with multitudes of invisible Bacteria germs. He is driven to such a doctrine, not only by his own confessions concerning Bacteria, but also by the microscopical evidence to which I have referred.

So that in explaining the results of any experiments made with the view of throwing light upon the cause of fermentation or putrefaction, it is especially necessary to bear in mind two considerations:—

I. That dust filtered from the atmosphere cannot be proved to include living Bacteria; though it is known to contain a multitude of organic particles which may be capable in the presence of water, in accordance with Liebig's hypothesis, of acting as ferments.

II. It must also be recollected that, in the opinion of many, Life represents a higher function which is displayed by certain kinds of organic matter; and that this higher function may be deteriorated or rendered non-existent by an amount of heat which might not be adequate to decompose the organic matter itself.

It is all the more necessary to call attention to these two considerations, because M. Pasteur invariably speaks as though it had been established that the air contains multitudes of living Bacteria, when, really, he had only proved that the air contains a number of corpuscles resembling spores of fungi, andc. And, as I have already intimated, the existence of spores of fungi in the atmosphere, however well established, is of little or no importance as an explanation of the cause of a very large number of fermentations. Their presence is even of still less importance, owing to the fact of the co-existence with these fungus-spores, of multitudes of organic fragments, which—in accordance with the views of Liebig, Gerhardt, and other chemists—are capable of acting as ferments. To this latter consideration M. Pasteur never even alludes when he speaks (loc. cit. p.) of his "ensemencements," and of other experiments which are equally, or even more, capable of being interpreted in accordance with Liebig's views than with his own.

Bearing these considerations in mind, we shall be in a better position to enquire into the real interpretation that may be given to many of M. Pasteur's results, and into the question as to how far the facts which he records are favourable to his own, or to the adverse doctrine concerning the causes of fermentation.

In the memoir so often alluded to on "The Organized Corpuscles which exist in the Atmosphere," M. Pasteur adduced various kinds of evidence, tending, as he thought, to show that the first Bacteria which make their appearance in putrefying or fermenting solutions, have been derived from living Bacteria or their "germs," which pre-existed in the atmosphere.

Some of the experiments by which he endeavoured to substantiate this position were of a very simple nature. Their narration attracted much attention at the time, as it was supposed that by their means M. Pasteur had—as he pro-fessed—con-clu-sive-ly shown the erroneousness of the views of those who believed in what was called "spontaneous generation." These experiments were soon repeated by other observers, who, using different fluids, obtained quite opposite results. Thus it became obvious to impartial critics, that whilst the means adopted by M. Pasteur might be adequate to check the processes of fermentation or putrefaction in certain fluids, they were quite powerless to effect this when many other fluids were employed.

These particular experiments, however, still seem to exercise a very great influence on the minds of many in this country, who are either unaware of, or disbelieve in, the possibility of obtaining opposite results.

The chapter in which M. Pasteur detailed these experiments is thus entitled:—"Another very simple method of demonstrating that all the organized products of Infusions (previously heated) owe their origin to the corpuscles which exist suspended in the Atmosphere." Whilst claiming to have already rigorously established the validity of this conclusion by the experiments described in previous chapters, M. Pasteur adds:—"If there remained the least doubt on this subject, in the mind of the reader, it would be dissipated by the experiments of which I am now about to speak." (p. .)

Sweetened yeast-water, urine, infusions of pear and of beetroot, were placed in flasks with long necks, variously drawn out and bent. The flasks were subsequently treated as follows. M. Pasteur says:—"I then raise the liquid to the boiling-point for several minutes until steam issues abundantly from the extremity of the drawn-out neck of the flask, which is permitted to remain open. I then allow the flask to cool. But, singular fact—and one well calculated to astonish every one acquainted with the delicacy of the experiments relating to what is called 'spontaneous generation'—the liquid of this flask will remain indefinitely without alteration. The flask may be handled without any fear, it may be transported from place to place, allowed to experience all the seasonal variations of temperature, and its liquid does not undergo the slightest alteration, whilst it preserves its odour and its taste." If, however, the neck of one of these flasks be broken off close to the flask itself, then, according to M. Pasteur, the previously unaltered fluid will, in a day or two, undergo the ordinary changes, and swarm with Bacteria and Mucedineæ.

"The great interest of this method is," M. Pasteur adds, "that it completes, unanswerably, the proof that the origin of life in infusions which have been raised to the boiling point, is solely due to the solid particles which are suspended in the air." He believes that any living things pre-existing in the fluid itself would be destroyed by the high temperature to which it had been raised; and that those contained in the air of the flask would also be destroyed, if not expelled, by the process of ebullition. Believing that the air is the source of germs from which Life is first developed in infusions, he thinks that what rapidly enters at first, on the cessation of ebullition, has its germs destroyed by contact with the almost boiling liquid; whilst the air which enters subsequently, and more slowly, is supposed to deposit its germs in the various flexures of the tubes, so that none are able to reach the fluid itself. Infusions, thus protected, do not undergo putrefaction, says M. Pasteur, because the access of pre-existing living things is necessary for the initiation of this change, and such access is prevented by the tortuous and bent neck of the flask.

Others say that some fluids submitted to the conditions mentioned, will undergo putrefactive changes, and that, therefore, these experiments of M. Pasteur are utterly incapable of settling the general question as to the cause of fermentation and putrefaction, and also that concerning the origin of Life. Although acknowledging a certain difficulty in explaining the results which are sometimes attained by this method, some of us would rather confess this than confidently offer explanations—as M. Pasteur did—which may in a short time be stultified by the results of other experiments with different fluids.

Having previously shown that living things could appear and multiply in such a flask as M. Pasteur describes—in any flask, in fact,—which had been hermetically sealed during the ebullition of a suitable fluid within; this was deemed to be a result so contradictory to the explanations of M. Pasteur, that it appeared needless to add my testimony, as I could have done, to that of M. Victor Meunier and others, as to the different results obtainable by operating, in M. Pasteur's fashion, with different fluids. It seemed to me that if organisms were to be procured in flasks from which air had been altogether expelled, it was useless still to urge the preservative virtues of any process of filtration of air—with the object of showing that living things in infusions derived their origin from atmospheric germs. Obviously, if there were no atmosphere, there could be no atmospheric germs present; and if living things were, nevertheless, developed under these exclusive circumstances, how could M. Pasteur or his disciples still expect to convince others that the first living things in infusions always proceeded from pre-existing atmospheric germs—even although it could be shown, that in many cases, when these were filtered off by flasks with narrow and tortuous necks, no living things were developed in such fluids. Granting to the full the truth of such facts, they could do nothing to establish the doctrine of the origin of infusorial life from pre-existing atmospheric germs, so long as it could also be shown that living things might be developed in boiled solutions to which air, instead of being filtered, was never allowed to enter at all.

It is not, therefore, because I think that some of the experiments which will subsequently be related afford any stronger or more direct support to my own conclusions, but because I think they may do this indirectly—by shaking the faith of many in some of the reasonings of M. Pasteur—that I am induced to give an account of them.

What has been hitherto said, also applies to the more recent statements concerning the efficacy of cotton-wool as an agent for filtering germs from the atmosphere. Prof. Huxley says he has never seen putrefaction or fermentation occur after certain organic fluids have been boiled for ten or fifteen minutes, if a good plug of cotton-wool has been inserted into the neck of the flask in which they are contained whilst ebullition is going on, and has, subsequently, been allowed to remain in the same situation. Using other or perhaps stronger fluids, however, I have found that such a method of proceeding is by no means adequate to stop the growth and development of organisms. And, also, even if it had been always efficacious—the reason adduced could not hold good, in the face of my other experiments, which had shown that a development of life might go on in cases where the air, which had been similarly driven out, was subsequently, in place of being filtered, prevented from gaining access to the fluid.

If germs derived from the air are the sole causes of putrefaction, then, surely, deprivation from air ought to be just as efficacious as any process of filtration of air—more especially when the filtration or the deprivation have a common starting point. And the mode of procedure, in both cases, is precisely the same up to a certain point. A fluid is boiled for a short time in order to kill the germs which may be within the flask, and to expel its previously contained air. At a certain stage of the ebullition, this may be arrested, if we have to do with a bent-neck flask, or one whose neck is plugged with cotton-wool, and no change, it is said, will subsequently take place in the contained fluid, because the air which enters is, by either of these means, filtered from its germs. But if, whilst ebullition continued, the neck of the flask had been hermetically sealed—so as altogether to prevent the re-ingress of air—and if the fluid, thus contained in vacuo, would nevertheless undergo fermentation, obviously the former explanation must be altogether shelved.

In the face of M. Pasteur's explanations, and those of Professor Huxley, these frequent positive results with fluids contained in vacuo are absolutely contradictory. There may naturally arise, therefore, a very grave doubt as to the validity of the explanation adduced by M. Pasteur, and adopted by Professor Huxley and others.

All these experiments to which I have been alluding are based upon the supposition (assented to by Pasteur and Huxley) that Bacteria which pre-existed in the solution would certainly be destroyed by its being raised for a few minutes to a temperature of ° F. This conclusion is, I believe, perfectly correct, and in support thereof I will adduce the following additional information.

Limits of 'Vital Resistance' to Heat displayed by Bacteria and Torulæ.

After stating elsewhere, that Vibriones are partly broken up or disintegrated by an exposure for a few minutes to a temperature of ° F. in an infusion which is being boiled, and also that, in all probability, the life of Bacteria would be destroyed by such a treatment, I made the following remarks:— "With reference to these organisms, however, one caution is necessary to be borne in mind by the experimenter. The movements of monads and Bacteria may be, and frequently are, of two kinds. The one variety does not differ in the least from the mere molecular or Brownian movement, which may be witnessed in similarly minute, not-living particles immersed in fluids. Whilst the other seems to be purely vital—that is, dependent upon their properties as living things. These vital movements are altogether different from the mere dancing oscillations which not-living particles display, as may be seen when the monad or Bacterium darts about over comparatively large areas, so as frequently to disappear from the field. After an infusion has been exposed for a second or two to the boiling temperature, these vital movements no longer occur, though almost all the monads and Bacteria may be seen to display the Brownian movement in a well-marked degree. They seem to be reduced by the shortest exposure to a temperature of ° C. to the condition of mere not-living particles, and then they become subjected to the unimpaired influence of the physical conditions which determine these movements." I now have various facts to add in confirmation of these conclusions, and in extension of our knowledge concerning the vital resistance to heat of Bacteria and Torulæ.

It would be a most important step if we could ascertain some means by which these primary movements of living Bacteria might be distinguished from the secondary, or communicated, movements of not-living particles. In many cases, organisms that are truly living may only exhibit very languid movements, which, as movements, are quite indistinguishable from those that the same Bacteria may display when they are really dead. Because the movements, therefore, are of this doubtful character, persons are apt, unfairly, to argue that the Bacteria which present them, are no more living than are the minute particles of carbon obtained from the flame of a lamp, which may exhibit similar movements. This, however, is a point of view which becomes obviously misleading if too much stress is laid upon it; and it is more especially so in this case, when those Bacteria which display the most characteristic sign of vitality—viz., "spontaneous" division or reproduction—do, at the time, almost always exhibit only the same languid movements. Mobility is, in fact, not an essential characteristic of living

Bacteria, whilst the occurrence of the act of reproduction is the most indubitable sign of their life. It should be remembered, therefore, that any Bacteria which are almost motionless, or which exhibit mere Brownian movements, may be living, whilst those which spontaneously divide and reproduce, are certainly alive—whatever may be the kind of movement they present.

In any particular case, however, can we decide whether Bacteria, that have been submitted to a given temperature, and which exhibit movements resembling those known as Brownian, are really dead or living? If the movements are primary, or dependent upon the inherent molecular activity of the organism itself, they ought, it might be argued, to continue when the molecules of the fluid are at rest; if, on the other hand, they are mere secondary or communicated movements, impressed upon the organisms as they would be upon any other similarly minute particles, by the molecular oscillations of the fluid in which they are contained, then the movements ought to grow less, and gradually cease, as the fluid approaches a state of molecular rest—if this be attainable. Following out this idea, some months ago, I first tested the correctness of the assumption by experimenting with fluids containing various kinds of not-living particles; such as carbon-particles from the flame of a lamp, or freshly precipitated baric sulphate. However perfect may have been the Brownian movements when portions of these fluids were first examined beneath a covering-glass, they always gradually diminished, after the specimen had been mounted by surrounding the covering-glass with some cement or varnish. Thus prepared, no evaporation could take place from the thin film of fluid, and after one, three, four, or more hours—the slide remaining undisturbed—most of the particles had subsided, and were found to have come to a state of rest. In order still further to test these views, I took an infusion of turnip, containing a multitude of Bacteria whose movements were of the languid description, and divided it into two portions. One of these portions was boiled for about a minute, whilst the other was not interfered with. Then, after the boiled solution had been cooled, a drop was taken from each and placed at some little distance from one another on the same glass slip; covering-glasses half an inch in diameter were laid on, and the superfluous fluid beneath each was removed by a piece of blotting-paper. When only the thinnest film of fluid was left, the covering-glasses were surrounded by a thick, quickly-drying cement. Examined with the microscope immediately afterwards, it was generally found that the Bacteria which had been boiled presented a shrunken and shrivelled aspect—whilst some of them were more or less disintegrated—though, as far as movement was concerned, there was little to distinguish that which they manifested, from that of their plumper-looking relatives which had not been boiled.

If the specimens were examined again after twenty-four or more hours, there was still very little difference perceptible between them, as regards their movements. And the same was the case when the specimens were examined after a lapse of some days or weeks. One important difference does, however, soon become obvious. The Bacteria which have not been boiled, undergo a most unmistakeable increase within their imprisoned habitat; whilst those which have been boiled, do not increase. The two films may be almost colourless at first (if the Bacteria are not very abundant), but after a few days, that composed of unboiled fluid begins to show an obvious and increasing cloudiness, which is never manifested by the other. Microscopical examination shows that this cloudiness is due to a proportionate increase in the number of Bacteria.

Is the continuance of the movements of the organisms which had been boiled attributable to their extreme lightness, and to the slight difference between their specific gravity and that of the fluid in which they are immersed? I soon became convinced that this was one, if not the chief reason, when I found that Bacteria which had been submitted to very much higher temperatures, behaved in precisely the same manner as those which had been merely boiled, and also that other particles which—though obviously dead—had a similar specific lightness, also continued to exhibit their Brownian movements for days and weeks. This was the case more especially with the minute fat particles in a mounted specimen of boiled milk, and also with very minute particles which were gradually precipitated from a hay infusion that had been heated to ° F. for four hours. Trials with many different substances, indeed, after a time convinced me that the most rapid cessation of Brownian movements in stationary films, occurred where the particles were heavy or large; and that the duration of the movement was more and more prolonged, as the particles experimented with, were lighter or more minute. So that, when we have to do with Bacteria, the minute oil globules of milk, or with other similarly light particles, the movements continue for an indefinite time, and are, in part, mere exponents of the molecular unrest of the fluid. They are always capable of being increased or renewed by the incidence of heat or other disturbing agencies.

In respect of the movements which they may exhibit, therefore, really living, though languid, Bacteria, cannot always be discriminated from dead Bacteria. Both may only display mere Brownian movements.

It becomes obvious, then, that in doubtful cases we ought not to rely very strongly upon the character of their movements, as evidence of the death of Bacteria—although these may frequently be of so extensive a nature as to render it not at all doubtful whether the Bacteria which display them are living. In the experiments which I am about to relate, we shall be able to pronounce that the Bacteria are living or dead, by reference to the continuance or cessation of their most essentially vital characteristic. If

Bacteria fail to multiply in a suitable fluid, and under suitable conditions, we have the best proof that can be obtained of their death.

Having made many experiments with solutions of ammonic tartrate and sodic phosphate, I have almost invariably observed that such solutions—when exposed to the air without having been boiled—become turbid in the course of a few days owing to the presence of myriads of Bacteria and Vibriones, with some Torulæ. These organisms seem to appear and multiply in such a solution almost as readily as they do in an organic infusion. On the other hand, having frequently boiled such solutions, and closed the flasks during ebullition, I have invariably found, on subsequent examination of these fluids, that whatever else may have been met with, Bacteria and Vibriones were always absent. The difference was most notable, and it seemed only intelligible on the supposition that any living Bacteria or dead ferments which may have pre-existed in the solution, were deprived of their virtues by the preliminary boiling. These experiments also seemed to show that such solutions, after having been boiled, and shut up in hermetically-sealed flasks, from which all air had been expelled, were quite incapable of giving birth to Bacteria. The unboiled fluid, exposed to the air, might have become turbid, because it was able to nourish any living Bacteria which it may have contained, or because it was capable of evolving these de novo, under the influence of dead ferments whose activity had not been destroyed by heat. Hence we have a fluid which is eminently suitable for testing the vital resistance of Bacteria,—one which, although quite capable of nourishing and favouring their reproduction, does not appear capable of evolving them, when, after previous ebullition, it is enclosed in a hermetically sealed flask from which all air has been expelled. Three flasks were half-filled with this solution. The neck of the first (a) was allowed to remain open, and no addition was made to the fluid. To the second (b), after it had been boiled and had become cool, was added half a minim of a similar saline solution, which had been previously exposed to the air, and which was quite turbid with Bacteria, Vibriones and Torulæ. From this flask—after its inoculation with the living organisms—the air was exhausted by means of an air-pump, and its neck was hermetically sealed during the ebullition of the fluid, without the flask and its contents having been exposed to a heat of more than ° F. The third flask (c) was similarly inoculated with living Bacteria, although its contents were boiled for ten minutes (at ° F.), and its neck was hermetically sealed during ebullition. The results were as follows:—the solution in the first flask (a), became turbid in four or five days; the solution in the second (b), became turbid after thirty-six hours; whilst that in the third flask (c), remained perfectly clear. This latter flask was opened on the twelfth day, whilst its contents were still clear, and on microscopical examination of the fluid no living Bacteria were to be found. This particular experiment was repeated three times with similarly negative results, although on two occasions the fluid was only boiled for one minute instead of ten.

It seemed, moreover, that by having recourse to experiments of the same kind, the exact degree of heat, which is fatal to Bacteria and Torulæ might be ascertained. I accordingly endeavoured to determine this point. Portions of the same saline solution, after having been boiled and allowed to cool, were similarly inoculated with a drop of very turbid fluid, containing hundreds of living Bacteria, Vibriones, and Torulæ. A drying apparatus was fixed to an air-pump, and the flask containing the inoculated fluid was securely connected with the former by means of a piece of tight india-rubber tubing, after its neck had been drawn out and narrowed, at about two inches from the extremity. The flask containing the inoculated fluid was then allowed to dip into a beaker holding water at ° F., in which a thermometer was immersed. The temperature of the fluid was maintained at this point for fifteen minutes, by means of a spirit lamp beneath the beaker. The air was then exhausted from the flask by means of the pump, till the fluid began to boil; ebullition was allowed to continue for a minute or two, so as to expel as much air as possible from the flask, and then, during its continuance, the narrowed neck of the flask was hermetically sealed by means of a spirit-lamp flame and a blowpipe. Other flasks were similarly prepared, except that they were exposed to successively higher degrees of heat—the fluid being boiled off, in different cases, at temperatures of °, °, °, °, and ° F. All the flasks being similarly inoculated with living Bacteria, Vibriones, and Torulæ, and similarly sealed during ebullition, they differed from one another only in respect to the degree of heat to which they had been submitted. Their bulbs were subsequently placed in a water bath, which during both day and night was maintained at a temperature of from ° to ° F. The results have been as follows:—The flasks whose contents had been heated to ° and ° F. respectively, began to exhibit a bluish tinge in the contained fluid after the first or second day; and after two or three more days, the fluid in each became quite turbid and opaque, owing to the presence and multiplication of myriads of Bacteria, Vibriones and Torulæ; the fluids in the flasks, however, which had been exposed to the higher temperature of °, °, °, and ° F., showed not the slightest trace of turbidity, and no diminution in the clearness of the fluid while they were kept under observation—that is, for a period of twelve or fourteen days. One kind of conclusion only is to be drawn from these experiments, the conditions of which were in every way similar, except as regards the degree of heat to which the inoculated fluids were subjected—seeing that the organisms were contained in a fluid, which had been proved to be eminently suitable for their growth and multiplication. If inoculated fluids which have been raised to ° and ° F. for ten minutes, are found in the course of a few days to become turbid, then, obviously, the organisms cannot have been killed by such exposure; whilst, if similar fluids, similarly inoculated, which have been raised to temperatures of °, °, °, and ° F. remain sterile, such sterility can only be explained by the supposition that the organisms have been killed by exposure to these temperatures.

Some of these experiments have been repeated several times with the same results. On three occasions, I have found the fluid speedily become turbid, which had only been exposed to ° F. for ten minutes, whilst on three other occasions I have found the inoculated fluid remain clear, after it had been exposed to a heat of ° F. for ten minutes.

In experimenting upon rather higher organisms, with which there is little difficulty in ascertaining, by microscopical examination, whether they are living or dead, I have found that an exposure even to the lower temperature of ° F. for five minutes, always suffices to destroy all signs of life in Vibrios, Amœbæ, Monads, Chlamydomonads, Euglenæ, Desmids, Vorticellæ, and all other Ciliated Infusoria which were observed, as well as in free Nematoids, Rotifers, and other organisms contained in the fluids which had been heated.

These results are quite in harmony with the observations and experiments of M. Pouchet and of Professor Wyman, as to the capability of resisting heat displayed by Vibriones and all kinds of ciliated infusoria. According to the former, the majority of ciliated infusoria are killed at, or even below, the temperature of ° F., whilst large Vibriones are all killed at a temperature of ° F. According to the observations of Professor Wyman, the motions of all ciliated infusoria are stopped at less than ° F., whilst Vibriones, taken from the most various sources, also seemed to be killed at temperatures between °–.° F. Similarly, we find Baron Liebig quite recently making the following remarks concerning a species of Torula:—"A temperature of ° C. [° F.] kills the yeast cells; after exposure to this temperature in water, they no longer undergo fermentation, and do not cause fermentation in a sugar solution. . . . In like manner, active fermentation in a saccharine liquid is stopped when the liquid is heated to ° C., and it does not recommence again on cooling the liquid."

That the organisms in question—being minute naked portions of living matter—should be killed by exposure to the influence of a fluid at these temperatures will perhaps not seem very improbable to those who have attempted to keep their fingers for any length of time in water heated to a similar extent. With watch in hand I immersed my fingers in one of the experimental beakers containing water at ° F., and found that, in spite of my desires, they were hastily withdrawn, after an exposure of less than five-and-twenty seconds.

Wishing to ascertain what difference there would be if the inoculated fluids were exposed for a very long time, instead of for ten minutes only, to certain temperatures, I prepared three flasks in the same manner—each containing some of the previously boiled solution, which, when cold, had been inoculated with living Bacteria, Vibriones, and Torulæ. These flasks and their contents were then submitted to the influence of the following conditions:—One of them was heated for a few minutes in a beaker containing water at ° F., and then by means of the air-pump a partial vacuum was procured, till the fluid began to boil. After the remainder of the air had been expelled by the ebullition of the fluid, the neck of the flask was hermetically sealed, and the flask itself was subsequently immersed in the water of the beaker, which was kept for four hours at a temperature between ° and

° F. The two other flasks similarly prepared were kept at a temperature of

°_

° F. for four hours. In two days, the fluid in the first flask became slightly turbid, whilst in two days more the turbidity was most marked. The fluid in the two other flasks which had been exposed to the temperature of

°_

° F. for four hours, remained quite clear and unaltered during the twelve days in which they were kept in the warm bath under observation. These experiments seem to show, therefore, that the prolongation of the period of exposure to four hours, suffices to lower the vital resistance to heat of Bacteria and Torulæ by

°_° F.

Such experiments would seem to be most important and crucial in their nature. They may be considered to settle the question as to the vital resistance of these particular Bacteria, whilst other evidence points conclusively in the direction that all Bacteria, whencesoever they have been derived, possess essentially similar vital endowments. Seeing also that the solutions have been inoculated with a drop of a fluid in which Bacteria, Vibriones, and Torulæ are multiplying rapidly, we must suppose that they are multiplying in their accustomed manner, as much by the known method of fission, as by any unknown and assumed method of reproduction. In such a fluid, at all events, there would be all the kinds of reproductive elements common to Bacteria, whether visible or invisible, and these would have been alike subjected to the influence of the same temperature. These experiments seem to show, therefore, that even if Bacteria do multiply by means of invisible gemmules as well as by the known process of fission, such invisible particles possess no higher power of resisting the destructive influence of heat than the parent Bacteria themselves possess. This result is, moreover, as I venture to think, in accordance with what might have been anticipated à priori. Bacteria seem to be composed of homogeneous living matter, and any gemmule, however minute, could only be a portion of such living matter, endowed with similar properties.

Extent to which boiled Fermentable Fluids may be preserved in Vessels with Bent Necks, or in those whose Necks are guarded by a Plug of Cotton-Wool.

Having thus satisfied ourselves as to the truth of the conclusion that Bacteria are killed when the fluid containing them is boiled (at ° F.), we are in a position to proceed with the inquiry as to the evidence which exists in respect to the statements made by M. Pasteur, Professor Huxley, and others, that fermentable fluids which have been boiled, will not undergo fermentation, either in vessels whose necks have been many times bent, or in those into whose necks a plug of cotton-wool has been inserted during the ebullition of their contained fluid. Organisms are not found in such cases, they say, because the "germs" from which the low organisms of infusions are usually produced, are arrested either in the flexures of the tube or in the cotton-wool. As I have before stated, however, it is obvious that if this explanation be the correct one, the preservation should be equally well marked in all cases—quite irrespectively of the amount of albumenoid or other nitrogenous material which may be contained in the fluid. Any exceptions to the rule should at once suggest doubts as to the validity of the explanation.

It was shown in by M. Victor Meunier that some fluids were preserved after having been boiled in a vessel of this kind, whilst others, submitted to the same treatment, speedily became turbid from the presence of Bacteria and other organisms. By these experiments he ascertained that strong infusions did frequently change, whilst weak ones might be preserved; and that even a strong infusion might be prevented from undergoing change if the period of ebullition were sufficiently prolonged.

The fluids most frequently employed by M. Pasteur were yeast-water, the same sweetened by sugar, urine, infusion of beetroot, and infusion of pear.

Taking urine as a fair example of such a fluid, I have found that the statements of M. Pasteur and of Professor Lister are perfectly correct. This fluid may generally remain for an indefinite period in such vessels without becoming turbid, or undergoing any apparent change. The same is generally found to be the case with an infusion of turnip, and occasionally an infusion of hay may be similarly prevented from undergoing fermentation. On the other hand, if the turnip-solution be neutralized by the addition of a little ammonic carbonate, or liquor potassæ; or, better still, if even half a grain of new cheese be added to the infusion before it is boiled, then I have found that the fluid speedily becomes turbid, owing to the appearance of multitudes of Bacteria. In an infusion to which a fragment of cheese had been added, I have seen a pellicle form in three days, which, on microscopical examination, proved to be composed of an aggregation of Bacteria, Vibriones, and Leptothrix filaments. A mixture of albuminous urine and turnip-infusion has also rapidly become turbid in a vessel of this kind owing to the appearance of multitudes of Bacteria, and so has a mixture containing one-third of healthy urine with two-thirds of infusion of turnip.

Other infusions have been boiled for ten minutes in a vessel with a horizontal neck two feet long, into which, during ebullition, a good plug of cotton-wool had been carefully pushed down for a depth of twelve or fourteen inches, and cautiously increased in quantity during the continuance of the ebullition; whilst immediately after the withdrawal of the heat, the plug was pressed closer, and all the outer unoccupied portion of the tube was rapidly filled up in the same manner.

Preserved in such a vessel, a specimen of urine remained unchanged; a hay-infusion also underwent no apparent alteration; whilst a very strong infusion of turnip became turbid in five days, and ultimately showed a large quantity of deposit.

Thus the rules laid down by Pasteur and others are not universal, and therefore, as I have previously pointed out, the explanation which he adduced of the preservation of those particular fluids which remained unchanged is at once rendered doubtful. More especially is there room for doubt on this subject when, as I have found, the result of the experiment can be, within certain limits, predicated beforehand, according to the nature of the fluid employed. If all organisms proceed from pre-existing germs, and these can be filtered from the air by a certain mechanical contrivance, then, if it be alleged that it is on account of such filtration that certain boiled fluids do not change, all fluids placed under these conditions ought, on this theory, to be similarly preserved. Exceptional cases cannot be accounted for on this hypothesis. To others, however, who say that organisms are capable of arising de novo, and that fermentation can be initiated without the agency of living things, the above facts appear quite natural. The more complex the nitrogenous or protein materials contained in a solution, the more is it fitted to undergo fermentative changes, which may be accompanied by the de novo origination of living things. Therefore the above results are just as compatible with the notions of M. Liebig and his school, as they are antagonistic to those of M. Pasteur. Certain fluids, it is found, do not undergo change; whilst other fluids, of a more complex description, will ferment under the influence of similar conditions. Prolonged ebullition also, by breaking up some of the more unstable compounds of a solution (those which most easily initiate these changes) will retard or prevent its fermentation.

The complete untenability of M. Pasteur's explanations are, however, best revealed by having recourse to a series of comparative experiments, in which portions of the same fluid are boiled for an equal length of time in vessels of different kinds, and are then subsequently submitted, in a water-bath, to the influence of the same temperature.

I have made many experiments of this kind with different solutions, some of which I will now record. Owing to the different behaviour of the same fluids under different conditions, we are enabled to draw some most important conclusions; and owing to the different behaviour of different fluids under these respective conditions, our attention is strongly drawn to other facts which ought considerably to influence our judgment as to the relative merits of the two doctrines concerning the cause of fermentation and putrefaction.

COMPARATIVE EXPERIMENTS.

In the following experiments, each fluid (unless a statement is made to the contrary) was boiled continuously for ten minutes, after having been placed in its flask. Then, with the neck either open, sealed, or plugged, the bulb of the flask was immersed in a water-bath maintained at a temperature of °–° F., during both day and night.

First Set of Experiments (I.–XV.).

a. Fluid exposed to Air in a Flask with a short Open Neck.

No. I.—Urine in twenty-four hours was still clear and free from deposit. In forty-four hours the fluid was very slightly turbid, and on microscopical examination Bacteria and Torulæ were found, though not in very great abundance. In sixty-eight hours the fluid was quite turbid.

No. II.—Hay Infusion in twenty-four hours was still clear. In forty-four hours the fluid was very turbid, and a drop on examination showed multitudes of Bacteria of different kinds, exhibiting languid movements. In sixty-eight hours the turbidity had become much more marked, and there was also a certain amount of sediment.

No. III.—Turnip Infusion in twenty-four hours showed a very slight degree of turbidity. A drop examined microscopically revealed a number of very minute, but very active, Bacteria. In forty-four hours the turbidity had become very well marked.

b. Fluid in contact with Ordinary Air and its Particles; Neck of Flask Sealed after the Fluid had become Cold.

No. IV.—Urine remained quite bright and clear during the fifteen days in which it was kept under observation in the water-bath.

No. V.—Hay Infusion after forty-four hours showed a well-marked turbidity. In sixty-eight hours there was an increase in the amount of turbidity, and also some sediment. During the next forty-eight hours turbidity and sediment gradually increased, whilst the colour of the fluid (originally that of port wine) became several shades lighter. Except that it grew still lighter in colour, and that the amount of sediment increased, it underwent no further obvious change during the fifteen days in which it remained in the bath.

No. VI.—Turnip Infusion underwent no change during the fifteen days in which it was kept in the bath under observation.

c. Fluid in a Flask with a Neck two feet long, and having Eight acute

Flexures.

No. VII.—Urine remained quite bright and clear during the fifteen days in which it was kept under observation in the water-bath.

No. VIII.—Hay Infusion remained bright and clear for twelve days. On the thirteenth day a very slight (almost inappreciable) sediment was seen, which scarcely underwent any obvious increase during the next eight days, though on the two following days (twenty-second and twenty-third) the turbidity became most obvious: much sediment was deposited, and the fluid assumed a much lighter colour. (On the twenty-second day the temperature of the bath was raised to ° F., for two or three hours.)

No. IX.—Turnip Infusion remained for four days without undergoing any apparent change. Its neck was then accidentally broken at the fourth joint— a certain amount of fluid still filling the third joint. In this condition the flask was allowed to remain in the water-bath, and the fluid continued quite unchanged in appearance for five days. It was then boiled for three minutes, and the neck of the flask was hermetically sealed whilst the fluid was boiling. The flask being re-immersed in water-bath, the fluid continued quite clear for thirteen days. Its neck was then carefully heated in the spirit-lamp flame till, when red-hot, the rapid inbending of the glass showed that the vacuum was still preserved. This being ascertained, the flask was, after a few minutes, replaced in the bath. The next day the temperature of the bath was allowed to go up to ° F. for three or four hours, and in the evening the fluid was observed to be very slightly turbid. In two days more (i.e., after sixteen days in vacuo) the turbidity was well marked, and when the fluid was examined microscopically it was found to contain an abundance of very languid Bacteria and Vibriones. On opening the flask there was an outrush of very fœtid gas, and the reaction of the fluid was acid.

d. Fluid in a Flask having a Neck two feet long, bent at right angles shortly above the bulb, and provided with a firm Plug of Cotton-Wool twelve inches in length.

No. X.—Urine remained quite bright and clear during the fifteen days in which it was kept under observation in the water-bath.

No. XI.—Hay Infusion showed a very slight amount of sediment after forty-four hours, which seemed to increase somewhat during the next three days. The fluid afterwards appeared to undergo no further change, though it remained in the warm water-bath for fifteen days.

No. XII.—Turnip Infusion in four days showed a well-marked turbidity, and also very many flakes of a broken pellicle.

e. Fluid (in vacuo) in a Flask, the Neck of which was hermetically Sealed by means of the Blowpipe Flame during Ebullition.

No. XIII.—Urine in forty-four hours showed a very slight amount of sediment. During the next two days the sediment very slightly increased, but was still small in amount. At the expiration of fifteen days, no further increase in the turbidity having taken place, the fluid was examined. The vacuum was still partially preserved, as evidenced by the rapid inbending of a portion of the neck of the flask after it had been carefully made red-hot. When opened, the odour of the fluid was stale, but not foetid, and its reaction was still faintly acid. On microscopical examination Bacteria and Torulæ were found in tolerable abundance.

No. XIV.—Hay Infusion in forty-four hours showed a very slight amount of turbidity. In sixty-eight hours the turbidity was most marked, and there was also a small amount of sediment. In another twenty-four hours it was noticed that the colour of the fluid had become much lighter, whilst the turbidity and sediment had increased. It subsequently continued in much the same state, and the flask was opened on the sixteenth day. The vacuum was found to be almost wholly impaired, whilst the odour of the fluid was sour, and not at all hay-like. On microscopical examination Bacteria, Vibriones, Leptothrix, and Torulæ, were found in abundance, and the former were very active.

No. XV.—Turnip Infusion after forty-eight hours showed a well-marked turbidity. In seventy-two hours the turbidity was more marked, and there was a slight amount of sediment. The turbidity also increased during the next twenty-four hours; though, after that, the infusion seemed to undergo no further change. The flask remained in the warm bath for fifteen days, when the fluid was examined. Its odour was not foetid, but was somewhat like that of baked turnip. Bacteria and Vibriones existed in abundance, though their movements were extremely languid.

Second Set of Experiments (XVI.–XXI.).

b. Fluid in contact with Ordinary Air and its Particles; Neck of Flask Sealed after the Fluid had become Cold.

No. XVI.—Simple Turnip Infusion in twenty-four hours had undergone no apparent change. In thirty-six hours there was slight turbidity, and in forty-eight hours this was most marked and uniform. When the flask was opened, after seventy-two hours, there was an outrush of very foetid gas; the reaction of the fluid was acid, and, when examined microscopically, it was found to contain multitudes of very languid Bacteria.

No. XVII.—Neutralized Infusion of Turnip +

gr. of Cheese, in thirty-six hours showed a well-marked pellicle. When the flask was opened, after seventy-two hours, there was a violent outrush of gas, though the fluid was still neutral. Portions of the thick pellicle were found, on microscopical examination, to be made up of Bacteria, Vibriones, and an abundance of long, interlaced Leptothrix filaments. Bacteria also existed abundantly in the fluid, though their movements were very languid.

c. Fluid in a Bent Neck Flask, having Eight acute Flexures.

No. XVIII.—Simple Turnip Infusion after forty-eight hours showed no change. It was kept in water-bath for twelve days, and during the whole of this time the fluid remained quite clear. The tube was then broken inch above the bulb (which was re-immersed in the bath), leaving the fluid exposed to the air through the straight open tube. The fluid at this time was odourless, and its re-action was still faintly acid.

The infusion remained thus exposed for six days without undergoing any apparent change. On the eighth day a very slight whitish sediment was noticed, which had increased in quantity by the tenth day, though there was still no trace of general turbidity. On the eleventh day some of the sediment was examined in a drop of the fluid, and it was found to be wholly composed of rather large Torulæ cells—the largest being about in diameter, though all the smaller sizes were abundantly represented. Not a single Bacterium or Vibrio could be detected, and the fluid was still quite odourless.

No. XIX.—Neutral Turnip Infusion + gr. of Cheese, showed no perceptible change in twenty-four hours, though in thirty-six hours there was a well-marked pellicle on the surface. When the neck of the flask was broken after seventy-two hours, the fluid was found to be very fœtid, whilst its re-action had become slightly acid. Portions of the pellicle were found to be made up by aggregations of Bacteria, Vibriones, and an abundance of Leptothrix filaments. The Bacteria all exhibited very languid movements.

e. Fluid (in vacuo) in a Flask which had been Sealed during Ebullition.

No. XX.—Simple Turnip Infusion in twenty-four hours showed a very slight amount of turbidity; in thirty-six hours this had increased, and in forty-eight hours there were multitudes of curdy flocculi floating in a tolerably clear fluid. The flask was opened after seventy-two hours, when there seemed to be only a very slight inrush of air. The odour of the fluid was somewhat fœtid, and its re-action was acid. There were multitudes of Bacteria and Vibriones, partly separate and partly aggregated (constituting the flocculi above mentioned). The separate Bacteria exhibited only very languid movements.

No. XXI.—Neutral Turnip Infusion + gr. of Cheese, showed a well-marked pellicle on its surface in twenty-four hours. In thirty-six hours the first pellicle had, in great part, sunk to the bottom of the flask, though its place on the surface was already taken by a new, though thin, scum-like layer. After seventy-two hours, the flask was opened; there was no fœtid odour of the fluid, and its re-action was still neutral. Examined microscopically the fluid showed an abundance of Bacteria, and also of short monilated filaments. There were, however, none of the ordinary kind of Vibriones, and no Leptothrix. All the Bacteria exhibited very languid movements.

Third Set of Experiments (XXII.–XXX.).

a. Fluid exposed to Air in a Flask with a short Open Neck.

No. XXII.—Urine in twenty-four hours showed no change; though in forty-six hours the turbidity was well marked. Examined microscopically it was found to contain an abundance of Bacteria.

b. Fluid in contact with Ordinary Air and its Particles; Neck of Flask Sealed after the Fluid had become Cold.

No. XXIII.—Urine in eighteen hours showed a distinct pellicle, though there was not much general turbidity. During the next few days the old pellicle fell to the bottom, and a new one formed.

c. Fluid in a Bent Neck Flask, having Eight acute Flexures.

No. XXIV.—Urine in forty-eight hours showed no change. After twelve days there was still no general turbidity, though there was a slight flocculent deposit of an uncertain nature. Two days afterwards the flask was broken, when the odour of the fluid was still found to resemble that of fresh urine, and its re-action was acid. The flocculi were made up of granular aggregations, in the midst of which were a few bodies closely resembling Torulæ, though they were somewhat doubtful in nature. Neither Bacteria nor Vibriones could be found. The flask, having a short open neck, was then replaced in the warm bath. In sixteen hours the whole fluid had become turbid; it was also slightly fœtid; and on microscopical examination it was found to be swarming with Bacteria, Vibriones, and Leptothrix.

No. XXV.—Turnip Infusion +

gr. of Cheese in forty-eight hours showed no change, though in seventy-two hours there was a well-marked pellicle, in which some bubbles of gas were engaged. After ninety-six hours the neck of the flask was broken; the fluid was found to be fœtid, and it had an acid re-action. On microscopical examination, a portion of the pellicle was seen to consist of multitudes of Bacteria, Vibriones, and jointed Leptothrix filaments.

No. XXVI.—Simple Turnip Infusion remained clear, and showed no appreciable change for seven days. On the eighth day a slight general turbidity of the fluid was noticed. On the ninth, the turbidity was rather more marked, though there was no trace of a pellicle; the neck of the flask having been broken, the fluid was found to be odourless and very faintly acid. On microscopical examination, multitudes of languid Bacteria of medium size were found, and also short monilated chains with from two to ten segments. There were no Vibriones, Leptothrix or Torulæ.

e. Fluid (in vacuo) in a Flask, Sealed during Ebullition.

No. XXVII.—Healthy Urine after twenty-four hours showed no change. After eleven days there was still no apparent change, though on the thirteenth a slight amount of flocculent sediment was noticed. This deposit increased in amount, very slowly, during the next fortnight; though afterwards the fluid seemed to undergo no further change, and did not become generally turbid.

No. XXVIII.—Healthy Urine (

) and Filtered Turnip Infusion (

) after forty-eight hours showed a very slight turbidity, which, however, became quite marked in another twenty-four hours.

No. XXIX.—Albuminous Urine (

) and filtered Turnip Infusion (

) after twenty-four hours, showed a slight turbidity, which became much more marked in forty-eight hours; whilst in seventy-two hours there was a considerable deposit at the bottom of the flask.

No. XXX.—Simple Turnip Infusion showed no change in forty-eight hours, though in seventy-two hours there was well-marked turbidity. The turbidity and sediment continued to increase for several days, and both were most marked on the tenth day, when the flask was opened. There was an outrush of gas, having an extremely fœtid odour. The fluid had an acid re-action, and when examined microscopically, multitudes of Bacteria, Vibriones and Leptothrix filaments were found—the movements of the Bacteria being very languid.

Fourth Set of Experiments (XXXI.–XXXVII.).

b. Fluid in contact with ordinary Air and its Particles; Flask Sealed after the Fluid had become Cold.

No. XXXI.—Healthy Urine remained in the warm bath for twenty-eight days without undergoing the least change.

No. XXXII.—Simple Turnip Infusion remained in the warm bath for twenty-eight days without undergoing any appreciable change. On breaking the neck of the flask, the fluid was found to be quite odourless. With its neck quite open, the flask was replaced in the water-bath. During the first forty-eight hours it underwent no apparent change, though at the end of seventy-two hours a slight general turbidity was noticeable, and an examination of a drop of the fluid (still odourless), showed a number of minute but very active Bacteria.

c. Fluid in a Bent-Neck Flask, having Eight acute Flexures.

No. XXXIII.—Simple Turnip Infusion showed no change after eight days' immersion in the warm bath. After eleven days, the fluid being still clear, the tube was broken just beyond the second bending from the bulb, and then the flask was re-immersed in the bath. After three days' exposure, the fluid being still clear, it was boiled in the flask for one minute, when it was noticed that the steam was quite odourless. The flask was then replaced in the water-bath, where it remained for twenty-two days (still with the neck open and broken just beyond its second bending) without showing any change. It was then submitted to examination; the fluid was found to be devoid of all odour, it had a slightly bitter taste, and its re-action was very faintly acid. On microscopical examination no living things were found; there were no Bacteria, no Vibriones, and no Torulæ, only some mere granules, a small amount of amorphous matter, and a few fibres.

No. XXXIV.—Turnip Infusion Neutralized by Ammonic Carbonate in forty-eight hours showed a slight turbidity, which slowly increased during the next two days. In two days more the turbidity was very great, and there was also a considerable amount of sediment. The fluid was then examined microscopically, and found to contain myriads of large but very languid Bacteria.

e. Fluid (in vacuo) in a Flask which had been Sealed during Ebullition.

No. XXXV.—Healthy Urine underwent no apparent change for the first twelve days, then (the bulk of the fluid still remaining clear and bright) small greyish white flocculi began to collect at the bottom of the flask, which very slowly increased in quantity during the succeeding twelve days. At the expiration of this time the flocculi were pretty numerous, though the fluid was otherwise bright. The vacuum was ascertained to be still good, and on breaking the flask, the fluid was found to have a slightly acid re-action, though no appreciable odour. When examined microscopically, the flocculi were seen to be made up for the most part of mere granular aggregations (simple, and not in the form of Bacteria). Small Torula cells, however, existed in some quantity; also a few necklace-like chains, and a comparatively small number of Bacteria, some of which were tolerably active.

No. XXXVI.—Simple Turnip Infusion after twenty-four hours showed no sign of change, though in thirty-six hours it was slightly turbid. On the fourth day the turbidity was well-marked and general, though there were no flake-like aggregations. When examined microscopically, the fluid was found to contain multitudes of Bacteria.

No. XXXVII.—Turnip Infusion, Neutralized by Ammonic Carbonate in twenty-four hours was decidedly turbid. In thirty-six hours the turbidity was more marked, and there was a slight sediment. By the end of forty-eight hours both turbidity and sediment had notably increased. On the fourth day, there was a moderately clear fluid, containing an abundance of curdy or flake-like masses. When the flask was opened, these were found to be made up principally by the aggregation of myriads of Bacteria.

Fifth Set of Experiments (XXXVIII.–XLVII.).

Fluids not boiled, but half-filling hermetically Sealed Flasks, containing Ordinary Air.

No. XXXVIII.—Turnip Infusion in ten hours showed a slight amount of turbidity. After forty-eight hours this was very well-marked: there was a thick pellicle on the surface, and, in addition, a small amount of deposit. On examination, the fluid and the pellicle were found to contain an abundance of Bacteria, Vibriones and Leptothrix filaments.

No. XXXIX.—Turnip Infusion +

of Carbolic Acid after eight days showed no appreciable alteration in appearance, no trace of pellicle or deposit. When examined microscopically, however, the fluid was found to contain some very minute Bacteria, though they were by no means abundant.

No. XL.—Hay Infusion had become quite turbid in twenty-four hours, and several shades lighter in colour. After forty-eight hours the colour of the infusion was still lighter; there was more turbidity, and some sediment. On microscopical examination, the fluid was found to contain an abundance of Bacteria, Vibriones and short Leptothrix filaments.

No. XLI.—Hay Infusion +

of Carbolic Acid showed no apparent change after forty-eight hours, and when examined microscopically it revealed no trace of Bacteria, or other organisms. The neck of the flask was then again closed. On the twelfth day the fluid had still undergone no change in appearance, and when examined microscopically, it still showed no trace of organisms, though the fluid was—as it had been at the time of the first examination—full of minute, undissolved particles of carbolic acid.

Fluids boiled for five minutes, and half-filling hermetically Sealed Flasks containing Ordinary Air.

No. XLII.—Hay Infusion, after forty-eight hours, showed no change, and continued to remain quite clear and free from deposit until the twelfth day, when it was examined microscopically. No organisms of any kind could be detected.

No. XLIII.—Hay Infusion +

part of Carbolic Acid showed no apparent change for the first five days, though, on the sixth day, a slight deposit was noticed at the bottom of the flask. The deposit had increased, and was well-marked by the twelfth day, when, on microscopical examination, there were found amongst the granular flakes of the deposit, Torulæ of several varieties of size and shape. Many were spherical, others ovoid, or having an elongated oat-like shape: some were of the ordinary colour, and others were brownish in tint. The variety was most striking. No Bacteria were seen, though there were multitudes of active particles which seemed to differ from the minute spherules of undissolved carbolic acid.

Fluids (in vacuo)—boiled for five minutes, and Flasks Sealed during Ebullition.

No. XLIV.—Turnip Infusion, in seventy-two hours, showed a slight turbidity, which gradually increased. On the eighth day there was a considerable quantity of flake-like sediment, and some amount of general turbidity. On the thirteenth day the vacuum was found to be still partly preserved. When the flask was opened the fluid was perceived to have a fœtid odour, and an acid re-action; and, on microscopical examination, multitudes of Bacteria and Vibriones were seen. In the flake-like aggregations also (made up almost wholly of these organisms) there were a number of large thick-walled spores; some already formed, and others in process of formation by coalescence.

No. XLV.—Turnip Infusion +

part of Carbolic Acid showed no increase of turbidity for the thirteen days during which it was kept under observation. Before the flask was opened it was ascertained that the vacuum was well preserved. The odour of the fluid was unaltered, and on microscopical examination no Bacteria, or other living things, were found.

No. XLVI.—Hay Infusion, after forty-eight hours, showed no change, though, in seventy-two hours, there was perceptible a very small amount of a dirty greyish deposit. By the fifth day the deposit had slightly increased, and on the seventh day there was a trace of turbidity in the fluid. It did not undergo much further change, so that, on the twelfth day, the flask was opened. The vacuum was found to have been very slightly impaired; the odour of the fluid was almost natural, and its re-action was slightly acid. On microscopical examination of the deposit, Bacteria, Vibriones, short Leptothrix filaments, and Torulæ, were found, though not in very great abundance.

No. XLVII.—Hay Infusion +

part of Carbolic Acid showed no apparent change for the first four days. On the fifth day there was a small quantity of powder-like sediment, and one dirty greyish-coloured flake. On the seventh day there were more small flakes at the bottom, and a slight general turbidity of the fluid. On the twelfth day, the turbidity and deposit having increased, the flask was opened—after it had been first ascertained that the vacuum had only been slightly impaired. The re-action of the fluid was still strongly acid. On microscopical examination of some of the deposit, there was found, amongst granular flakes and aggregations, a large number of Torulæ cells, of most various shapes and sizes; also in the midst of the granule heaps many large, rounded or ovoidal, densely granular nucleated bodies, whose average size was

" in diameter, though there were many of them much larger, and others even less than half this size. Intertwined amongst the granular matter also were a large number of algoid-looking filaments,

in diameter, containing segmented protoplasmic contents. There were also in the fluid itself a number of medium-size, unsegmented Bacteria, whose movements were somewhat languid.

Sixth Set of Experiments (XLVIII.–LXV.).

Ammoniacal Solutions, unboiled, and exposed to Ordinary Air in a Corked Bottle. (Temp. °–° F.)

No. XLVIII.—Ammonic Acetate Solution.—On the tenth day the fluid was still quite clear, and free from sediment.

No. XLIX.—Ammonic Oxalate Solution.—On the tenth day there was no distinct opalescence of the fluid, but a well-marked whitish flocculent deposit. On microscopical examination no Bacteria were found in the fluid, and the deposit was made up by an aggregation of blackish and colourless granules, mixed with a few crystals and a very few Torula cells—all being held together by a sort of mucoid matrix. In the midst of this matter were found two or three very small, much branched, mycelial tufts of a fungus-growth.

No. L.—Ammonic Carbonate Solution.—On the tenth day the fluid showed a very faint opalescence, with a small amount of deposit, and a partial non-coherent scum on the surface, which, on microscopical examination, was found to be composed partly of amorphous granules, and partly of minute Bacteria, mixed with small necklace-like organisms. The fluid itself contained, in suspension, a few small and sluggish Bacteria, with a minute Torula cell here and there.

No. LI.—Ammonic Tartrate Solution after twenty-four hours showed the faintest opalescence of the fluid; in forty-eight hours there was a bluish-white turbidity, and in seventy-two hours the turbidity was well marked. When examined microscopically the fluid was found to contain multitudes of very active Bacteria. On the thirteenth day the turbidity was not so well marked, though there was a very thin pellicle on the surface, and also the dirty-looking crumpled remains of another pellicle at the bottom, which, on examination, was found to be composed of an aggregation of Bacteria. The pellicle on the surface was very thin, and composed only of a single layer of Bacteria. In the fluid itself many Bacteria were seen, of medium size, and mostly sluggish in movement, though a few of them exhibited very active rotatory movements. No Vibriones, Leptothrix, or Torulæ, were found.

No. LII.—Ammonic Tartrate and Sodic Phosphate Solution after twenty-four hours showed the faintest opalescence; in forty-eight hours there was a bluish-white turbidity, which, in seventy-two hours, had become more marked. When examined microscopically multitudes of Bacteria were found whose movements were very sluggish. On the thirteenth day there was a well-marked whitish turbidity, due to Bacteria and Vibriones, a slight amount of deposit, and a firm pellicle which was found to be composed, almost wholly, of long unjointed Vibriones and unsegmented Leptothrix filaments, all of which, when separate, exhibited the most distinct eel-like movements, accompanied by an actual progression from place to place.

Ammoniacal Solutions, unboiled, and exposed to Air in a Corked Bottle, after Inoculation with a Drop of Fluid containing living Bacteria and Torulæ. (Temp. °–° F.)

No. LIII.—Ammonic Acetate Solution after twenty-four hours was faintly opalescent, and in forty-eight hours showed a very slight bluish tint. In seventy-two hours it was in the same state, and, on microscopical examination, the fluid showed no distinct Bacteria or other living things, though there were a number of very minute particles distributed, singly or in small groups, throughout the fluid. On the thirteenth day there was no change in appearance, except that the sediment had somewhat increased in amount. Still, no Bacteria could be found in the fluid or the sediment,— only the above-mentioned particles, and a few somewhat larger, which resembled very minute Torulæ. Amongst the sediment, however, there were two or three very small mycelial tufts of a developing fungus.

No. LIV.—Ammonic Oxalate Solution.—On the eighth day the fluid showed a very faint opalescence, though there was a well-marked, greyish, flocculent deposit, which was found to be composed of an aggregation of colourless and blackish granules, of a multitude of minute crystalline particles (mostly diamond-shaped), and some rounded or ovoidal, thick-walled, spore-like bodies; amongst which, and enveloped in part by them, were several mycelial tufts of a fungus. A number of minute Bacteria were found distributed throughout the fluid, and also a quantity of minute star-like bodies (crystalline), about

in diameter.

No. LV.—Ammonic Carbonate Solution.—On the eighth day the fluid showed a very faint opalescence, and a slight deposit, which was found to be composed principally of amorphous granules. Distributed through the fluid were some small and sluggish Bacteria, though no other organisms were seen.

No. LVI.—Ammonic Tartrate Solution.—After twenty-four hours the fluid showed the faintest opalescence, and in forty-eight hours there was a slight bluish-white turbidity. In seventy-two hours the turbidity was well marked, and there was a very thin pellicle on the surface. When examined microscopically the fluid was found to contain multitudes of very active Bacteria, and the pellicle was also composed of an aggregation of Bacteria. On the thirteenth day the opacity had somewhat increased; there was also a well-marked pellicle, and an obvious deposit. The pellicle was found to be composed of Bacteria, and in the fluid there were multitudes of medium-size Bacteria and Vibriones, with here and there a small Torula cell.

Ammoniacal Solutions (in vacuo) in Flasks which were hermetically Sealed during Ebullition of their Fluids at a Temperature of ° F. (Subsequently exposed in water-bath to a Temperature of °–° F.).

No. LVII.—Ammonic Tartrate Solution after sixty hours showed a slight sediment, with bluish flakes attached to sides of flask. In eighty-four hours there was a general bluish opalescence, and on microscopical examination the fluid was found to contain multitudes of Bacteria.

No. LVIII.—Ammonic Tartrate and Sodic Phosphate Solution.—After sixty hours there was a slight general bluish opalescence. In eighty-four hours the general opalescence was not more marked, but there were many flake-like aggregations in the fluid, which on microscopical examination were found to be aggregations of Bacteria.

Ammoniacal Solutions boiled (at ° F.), and exposed to Air in Flasks whose Open Necks were only loosely covered with Paper Caps: subsequent Inoculation. (Temp. °–° F.).

No. LIX.—Ammonic Tartrate Solution.—The fluid remained quite clear, and free from all trace of turbidity up to the ninth day, when it was inoculated with some living Bacteria. In fifty hours after the inoculation there was a very faint opalescence of the fluid, which, in another hours, had become much more marked. On microscopical examination it was found to contain multitudes of Bacteria.

No. LX.—Ammonic Tartrate and Sodic Phosphate Solution.—After four days the fluid was still quite clear. In seven days no trace of general turbidity, though there was a minute dirty-grey aggregation about " in diameter at the bottom of the flask. On the sixteenth day the grey aggregation had very slightly increased in size, though the fluid above was still perfectly clear. The grey mass was removed by a small pipette, and, on microscopical examination, it was found to be composed of an aggregation of minute extraneous fibres, mixed with blackish particles and amorphous granular matter, in which were growing many Torula-cells in all stages of development, and also a minute mycelium composed of branched Leptothrix-like fibres. The clear fluid was then inoculated with some living Bacteria, and the bulb of the flask was replaced in the warm bath. After fifty hours the solution showed a bluish turbidity, which, in thirty-six hours more, had increased to a well-marked whitish opacity, and when examined, the fluid was found to be swarming with active Bacteria.

Solutions of Ammonic Tartrate and Sodic Phosphate were heated, in their respective Flasks, for Fifteen Minutes to the Temperatures mentioned below. The Necks of the Flasks were afterwards loosely covered with Paper Caps, whilst the Bulbs were immersed in a Water-Bath kept at a Temperature of °–° F.

No. LXI.—
Solution heated to
° F.

No. LXII.
° F.

No. LXIII.
° F.

No. LXIV.
° F.

No. LXV.

° F.

All these solutions remained quite clear and free from any trace of general turbidity for ten days. Each fluid was then inoculated with some living Bacteria, and in the space of thirty-six to seventy-two hours, all had become more or less obviously turbid, and on microscopical examination this turbidity was found in each case to be almost wholly due to the presence of multitudes of Bacteria.

Interpretation of the Experiments: Conclusions as to the Cause of Fermentation, and as to the Occurrence of Archebiosis.

These experiments seem to show quite conclusively that M. Pasteur's explanations are altogether inadequate to account for the occasional preservation of boiled fluids in bent-neck flasks. They show that the preservation, far from being universal, is only occasional, and that preservation or non-preservation of different fluids is almost wholly dependent upon their nature. They lend no countenance, moreover, to his particular theory, that fermentation cannot be initiated without the agency of living ferments,—they are, on the contrary, wholly opposed to this restriction.

The plug of cotton-wool, or the narrow and bent tube may, it is true, protect the boiled fluid from subsequent contact with living "germs"; but that the fluids do not undergo change on account of such deprivation cannot be safely affirmed, when the same means would also filter from the fluid some of the multitudinous particles of organic matter (dead), which the air undoubtedly contains, and which may act as ferments. It must be remembered that the main object of M. Pasteur's investigation was to determine whether fermentation took place under the agency of mere dead nitrogenous matter, as Liebig and others affirm; or whether it is only initiated by living organisms, as he himself supposes. Obviously, therefore, the same filtration which purified the air from any living organisms would filter from it its nitrogenous particles, which are the other possible ferments: so that no conclusion could be drawn from such experiments, more favourable to the one than to the other of these hypotheses. All that could have been safely affirmed was, that by boiling the fluid, and then protecting it from subsequent contact with everything that could act as a ferment, fermentation would not take place.

Even this however,—as the preceding experiments fully show—cannot be truly affirmed to be a general rule. Some infusions do undergo change, notwithstanding this treatment and deprivation, whilst others do not: that is to say, some still preserve a first degree of fermentability even after boiling, whilst others are reduced by this process to the second degree of fermentability. These latter are unable to initiate changes by virtue of their own inherent instability; molecular re-arrangements require to be set on foot in them by contact with some more unstable substance, which is itself undergoing change.

That such is the correct explanation of the reason why some fluids do not ferment in bent-neck flasks, seems obvious from the discordant results obtained in many other experiments, after the free admission of uncalcined air to the fluids which had been boiled. The fluids were deprived of their virtues in some cases by the heat to which they had been subjected, so that whether they underwent change or not, may have depended upon the accidental presence or absence, in the air which was subsequently admitted to the fluid, of some unheated organic fragments, capable of initiating fermentative changes. If germs were as omnipresent as they have been represented to be, such fluids ought always to have undergone fermentation.

Whilst I have found that any given fluid, whose strength is about equal on different occasions, acts in a definite manner when the flask is hermetically sealed after expulsion of all its air and during the continuance of ebullition; and, whilst a like definite result can generally be obtained, when calcined air is admitted to the boiled fluid before the vessel is hermetically sealed; it is found, on the other hand, that the result is in no way predicable when uncalcined air is admitted. Sometimes fermentation takes place, and sometimes in other flasks—sealed at the same time, and subsequently placed under the same conditions—no change whatever occurs. My own experience in this respect accords perfectly with that of M. Pasteur. He, however, at once came to the conclusion that the only inference from such facts was that "germs" are not so universally distributed as they had been supposed to be by Bonnet and Spallanzani. The unprejudiced inquirer, however, will perceive that M. Pasteur was entitled to come to no such conclusion concerning germs which was not equally applicable to minute fragments or débris of organic matter floating in the air. And, similarly, the evidence which he adduces with regard to the diminution in the number of the fertile flasks when they were filled with some of the still air of the caves of the observatory, or else with some from the peaks of the Jura, far away from the haunts of men, had no bearing upon the distribution of germs which was not equally applicable to that of dead organic particles. Such evidence, therefore, was valueless for settling between the rival doctrines of fermentation—it could not possibly help us to decide whether living or dead ferments were necessary. Dead organic particles would sink in still air in the same manner as living organisms; and similarly, dead organic particles, have been shown to be less and less numerous in the atmosphere in proportion to the elevation obtained. In these latter experiments M. Pasteur made use of yeast-water (alone or sweetened), and of urine—all three of them fluids, which, after having been boiled, are apt to possess only the second degree of fermentability. Shortly afterwards, M. Pouchet, in concert with MM. Joly and Musset, repeated these experiments, with the sole difference that they employed strong infusions of hay—which experiment has shown almost invariably to possess the first degree of fermentability. And seeing that all their flasks, after a time, yielded organisms from whatever mountain elevation the air had been taken—the combined evidence tends strongly against the view of M. Pasteur. Since the germs in the fluids and in the flasks, in each set of experiments, had been previously destroyed by ebullition, and since in each set, also, air of the same character had been admitted to the boiled fluids, the different results seemed to show that fermentation or non-fermentation, in such cases, depends wholly upon the quality of the fluids employed.

Other evidence which is so much vaunted by M. Pasteur and his supporters, as to the possibility of inducing fertility in previously sterile flasks, by the addition of a portion of asbestos containing the solid particles filtered from the atmosphere, is also equally valueless for confirming the proposition that fermentation is only capable of being initiated by living ferments. The same asbestos which may contain living spores or organisms ("germs"), does undoubtedly contain many decomposable particles and fragments of organic matter. The previously barren solution may therefore be rendered fertile by the mere addition of those portions of unstable organic matter, whose molecular mobility has not been impaired by the agency of heat, and which are therefore capable of initiating fermentations. This view is strengthened, as M. Pouchet has pointed out, by the fact that in these cases, instead of meeting some of the various kinds of organisms which are supposed to have representatives in the air, and whose spores or ova may be supposed to have been sown, it is often merely Bacteria which are encountered,—differing in no respect from those that may present themselves in a somewhat similar infusion, which has undergone change in a closed flask without any such hypothetical sowing of living spores or germs. It is more especially important to bear in mind this consideration, when portions of organic matter can always be easily demonstrated amongst such atmospheric dust; whilst living Bacteria, or other organisms, such as are first produced as a result of the supposed sowing of spores, either cannot be demonstrated, or would seem, from other evidence, to be at least very sparingly distributed.

The fact revealed by M. Pasteur, that some fluids remain unchanged for an indefinite period, after having been boiled in flasks, with long and bent necks, is easily explicable in accordance with the physical theory of fermentation, and now that it has been thoroughly proved that other fluids—submitted to precisely similar conditions—do nevertheless undergo fermentation, this fresh fact is just as completely adverse to the explanations and views of M. Pasteur, as it is thoroughly harmonious with the doctrines of Baron Liebig. The fluids which are capable of being preserved—generally not presenting a high degree of fermentability—do not undergo change, at ordinary atmospheric pressure, after having been boiled, unless they are brought into contact either with some pre-existing living things or with some unaltered organic particles from the atmosphere. Neither of these, however, can gain access to the fluid, in such a vessel; because all the air which enters, after the first inrush into the still almost boiling fluid, has to pass, more or less slowly, through the numerous flexures of the narrow neck of the flask and the two or three strata of fluid which always remain therein.

Some of the fluids which do not undergo change in these bent-neck vessels are, however, by no means notable for possessing a low degree of fermentability. This is the case, for instance, with infusions of turnip, which, under other conditions, have been found to be most prone to undergo fermentation. And, I have found in several cases in which such an infusion had been exposed in a bent-neck vessel, and had remained unchanged for twelve or fourteen days (even though subjected to a temperature of $°-°$ F.), that if the neck of the flask were then broken shortly above the bulb, the solution would still continue without alteration for a week, ten days, a fortnight, or even more—although freely exposed to the air, and therefore to the access of any living germs which might be floating about in the atmosphere.

The views hitherto expressed with reference to the causes of fermentation and putrefaction, and to the interpretations which M. Pasteur's experiments are capable of receiving, seem to derive all the additional support that can be needed, from the results of my own experiments with boiled fluids, in sealed flasks, from which all air had been expelled.

Some of the same fluid being taken and divided into three parts, each portion is placed in a separate flask, in which it is boiled for a period of ten minutes. One of the flasks (A) is provided with a long and bent neck, so that the air which re-enters is deprived of its germs and organic particles; another (B) has only a short neck, and to this, the access of germs and organic particles is freely permitted till the fluid has become cool, and then the neck of the flask is hermetically sealed; whilst the last (C) is sealed during ebullition, after all air has been expelled. Now, if Pasteur's theory of fermentation, and the prevalent notions concerning the universal distribution of "germs" throughout the atmosphere were true, it might be expected that the fluid in B would always rapidly change; that that in A would always remain pure; and that the fluid in C would, similarly, undergo no alteration. The facts, however, are quite the reverse: if a strong turnip infusion be employed, the fluid in A will almost always remain unchanged; that in B will sometimes rapidly change, and at other times will remain quite pure; whilst that in C will almost invariably become turbid in from two to six days. But, even if it were not the case that some fluids, different from those used by M. Pasteur, will almost invariably undergo change in bent-neck vessels, M. Pasteur's explanation of the cause of the preservation would have been altogether upset by the fact that some of the very fluids which remain pure in the bent-neck apparatus will become fœtid if shut up in vacuo. It is, therefore, of course useless to talk of a particular boiled fluid having been saved from putrefaction on the ground that the living atmospheric germs (whose presence is supposed to be necessary for the initiation of such a process) have been altogether filtered from the re-entering air, when some of the same fluid will putrefy, if placed under different conditions by which it is freed both from the influence of the atmosphere and from its germs—i.e., when, instead of filtering the re-entering air, no air is permitted to enter. Germs and atmospheric particles being equally got rid of in both sets of cases, the great difference between them is that the weight of the atmosphere is also got rid of in my experiments—the fluids being contained in vacuo. Now it has been ascertained by Mr. Sorby, that pressure undoubtedly influences "chemical changes taking place slowly, and therefore, probably due to weak or nearly counterbalanced affinities;" and he also states, in the Bakerian Lecture for , that "a considerable number of facts have been described, showing that pressure will more or less influence such chemical actions as are accompanied by an evolution of gas, so that it may cause a compound to be permanent, which otherwise would be decomposed." If increase of pressure retards, a diminution of pressure will facilitate such chemical changes, so that one can only explain the results which I have obtained, on the ground that many boiled fluids, which will not undergo change when protected from the influence of atmospheric particles (living or not living) at the same time that they are subjected to ordinary or increased pressure, will, on the contrary, pass through such changes when pressure is removed, and the fluids are preserved in vacuo. It is not pretended that this is a rule applicable to all organic fluids—far from it. Diminution of pressure does seem, however, to be a very potential cause of change in some fluids. The extent to which changes of a fermentative character can progress in the absence of atmospheric oxygen, is also evidently subject to much variation, in accordance with the nature of the dissolved fermentable substances.

These facts are not so new and exceptional, however, as they may at first sight appear. It has been long known that a boiled fluid extremely prone to change, will not yield infusoria if the vessel in which it is contained is filled with the fluid. Burdach says:—"Gruithuisen a reconnu que des infusions, même très fécondes d'ailleurs, celles du foin par exemple, ne donnaient point d'infusoires dans des flacons de verre dont le bouchon était assez enfoncé pour toucher à la surface de l'eau." On the other hand, no experiments with which I am acquainted, in which heated fluids and calcined air have been shut up in closed flasks, have yielded so many positive results as those of Professor Wyman of Cambridge, U.S.,—and his were performed under precisely the reverse conditions. Large flasks were used, and only

□ –□

of their bulk was filled by the experimental fluids. Gruithuisen's results were explained by Burdach on the ground that a certain amount of air was necessary, and it was also with the view of subjecting his fluids to as large an amount of air (calcined) as possible, that Professor Wyman employed small quantities of fluids in large flasks. These views were dictated by the chemical doctrines of Gay-Lussac and others, to the effect that the oxygen of the air is the initiator or primum movens of fermentative changes.

Now, without doubting in the least that in some instances this may be the case, it seems to me quite obvious, from my own experiments, that a different interpretation may be given of Gruithuisen's results—which I have myself verified,—and of the fact that meats and vegetables will often remain unchanged for years after having been heated in closed tins from which all air has been expelled.

If we ponder only upon the fact that certain fluids, in contact with a very small quantity of air, in an hermetically-closed vessel, will not undergo change; though these same fluids will change when exposed to a much larger quantity of calcined air, there may be strong reason for coming to an opinion similar to that of Gay-Lussac. When, however, it is also ascertained that provisions which have been subjected to the long-continued influence of heat, do not undergo change in closed vessels, if all air has been expelled from the very small space above the level of the provisions; though many organic infusions will putrefy if they occupy only one half, or less, of a hermetically closed vessel from which all air has been similarly expelled by ebullition of its fluid contents, it is impossible that the same explanation can hold good. And at the same time another interpretation is suggested for the first set of facts.

The last-mentioned experiments prove that fermentation can take place in vacuo, when the conditions are more favourable than those which present themselves within the almost full tins containing provisions. The change in these latter cases cannot (in the great majority of instances) proceed far, because there is no adequate space into which residual gases may be emitted. When this emission (which is almost always one of the accompaniments of a fermentative change) has taken place to a slight extent, the meats are in the very best condition for preservation. There is an utter absence of light, an absence of free oxygen, and also an absence of that diminished pressure which my experiments seem to show is favourable to the promotion of many kinds of fermentative change. So that if fermentation does not take place in a closed flask which is full of a boiled infusion of hay, it may be owing to the fact that there is no space for the residual gases, and that undue pressure retards many fermentative changes.

This is also perfectly compatible with the other fact that the same kind of fluid will undergo change when a small quantity of it is contained in a comparatively large flask—owing to there being, in such a case, plenty of room for residual gases to be effused, before that undue amount of pressure is brought about, in the presence of which such a fluid will no longer ferment or putrefy. Fluids, therefore, whose putrefaction is hindered by increased pressure and favoured by diminution of pressure, may be placed under conditions which are successively more favourable than the last, by putting a gradually smaller and smaller quantity of fluid into a flask, to which calcined air is admitted, and, better still—if the stimulus of oxygen is not absolutely needed in order to incite fermentation in the fluid employed—by only half filling the flask, and procuring a more and more perfect vacuum.

In accordance with the doctrines of Baron Liebig, therefore, my experiments, as well as those of many other investigators, tend to show that fermentative and putrefactive changes are merely processes of chemical re-arrangement, which frequently take place—as it were "spontaneously"— owing to the inherent instability of certain nitrogenous compounds in the presence of free oxygen. My experiments have, however, also revealed the additional fact that, under the combined influence of heat and diminished pressure, some fluids will undergo fermentation even in closed vessels, from which all air has been expelled. They lend no support to the idea that the air is so thickly laden with living germs as some would have us suppose; and in view of the mass of positive information now in our possession concerning the degree of heat which suffices to kill the lowest organisms, they also, as I think, entitle us to come to the conclusion that such organisms are (as the microscopical evidence might lead us to believe) really capable of being evolved de novo. These lowest organisms are, in fact, to be regarded as occasional concomitant products, rather than as invariable or necessary causes of all fermentative changes.

It would thus appear that specks of living matter may be born in suitable fluids, just as specks of crystalline matter may arise in other fluids. Both processes are really alike inexplicable—both products are similarly the results of the operation of inscrutable natural laws, and what seem to be inherent molecular affinities. The properties of living matter, just as much as the properties of crystalline matter, are dependent upon the number, kind, and mode of collocation of the atoms and molecules entering into its composition. There is no more reason for a belief in the existence of a special "vital force," than there is for a similar belief in the existence of a special "crystalline force." The ultimate elements of living matter are in all probability highly complex, whilst those of crystalline matter are comparatively simple. Living matter develops into Organisms of different kinds, whilst crystalline matter grows into Crystals of diverse shapes. The greater modifiability of living matter, and the reproductive property by which it is essentially distinguished from crystalline matter, seem both alike referable to the great molecular complexity and mobility of the former. Crystals are statical, whilst organisms are dynamical aggregates, though the evolution of both, marked by their peculiar characteristics, may be regarded as visible expressions testifying to the existence of one all-pervading Power "Whose dwelling is the light of setting suns,

And the round ocean, and the living air,

And the blue sky, and in the mind of man:

A motion and a spirit that impels

All thinking things, all objects of all thought,

And rolls through all things."

www.ingramcontent.com/pod-product-compliance
Lightning Source LLC
Chambersburg PA
CBHW071637170526
45166CB00003B/1359